A PLUME BOOK

SMARTER

DAN HURLEY is an award-winning science journalist whose 2012 feature in the *New York Times Magazine*, "Can You Make Yourself Smarter?" was one of the magazine's most-read articles of the year. In 2013 he published another article for the magazine, "Jumper Cables for the Mind," describing his experience with transcranial direct-current stimulation. He has written on the science of increasing fluid intelligence for the *Washington Post* and *Neurology*, and is featured in the 2013 PBS documentary *Smarter Brains*. His books have been excerpted in *Wired* and *Discover* magazine. Hurley has written nearly two dozen science articles for the *New York Times* since 2005.

Praise for *Smarter*

"*Smarter* is an essential read. It's a riveting look at the birth of a new science as well as a user's manual for anyone who wants to be better at solving problems, learning new things, and coming up with creative ideas."

—Daniel H. Pink, author of *Drive* and *A Whole New Mind*

"Hurley captures the history and mystery of intelligence, but, most of all, the exciting new science of intellectual growth. This may be the most important revolution of our time!"

—Carol Dweck, author of *Mindset: The New Psychology of Success*

"Dan Hurley isolates just what cognitive exercise boosts intelligence. Anyone who doubts that environment can make a real difference to cognition should start with this book."

—James R. Flynn, author of *What Is Intelligence?*

"Hurley's book shows that you can indeed make yourself (and your children) smarter."

—Richard E. Nisbett, PhD, author of *Intelligence and How to Get It*

"Filled with beautifully explained science, *Smarter* is engaging and inspiring, offering much-needed hope to those of us whose smarts seem to be declining. *Smarter*, in fact, is that rare thing: enjoyable reading that can also improve your life."

—Gretchen Reynolds, author of *The First 20 Minutes*

Smarter

The New Science of Building Brain Power

Dan Hurley

A PLUME BOOK

PLUME
Published by the Penguin Group
Penguin Group (USA) LLC
375 Hudson Street
New York, New York 10014

USA | Canada | UK | Ireland | Australia | New Zealand | India | South Africa | China
penguin.com
A Penguin Random House Company

First published in the United States of America by Hudson Street Press, a member of Penguin Group (USA) LLC, 2014
First Plume Printing 2015

Portions of this book were first published in slightly different form in the *New York Times Magazine* as "A Drug for Down Syndrome" on July 29, 2011, and "Can You Make Yourself Smarter?" on April 18, 2012; and in the Education Living section of the *New York Times* as "The Brain Trainers" on October 31, 2012.

 REGISTERED TRADEMARK—MARCA REGISTRADA

ISBN 978-0-14-218165-2 (pbk.)

Printed in the United States of America
10 9 8 7 6 5 4 3 2 1

Set in Adobe Garamond Pro

For my big brothers and little sister up in Maine:
John, Mike, Dave, Pat, and Eileen

We know what we are, but know not what we may be.

—Shakespeare, *Hamlet*

CONTENTS

INTRODUCTION

Danny and Julie Vizcaino, brother and sister, were born and raised in a poor neighborhood of Modesto, California, she in 1981, he in 1983. Their parents, immigrants from Mexico without high school diplomas, were typical of the local population: their mother worked in a canning factory, and their father worked in construction until he died in an accident when the kids were young. With an older brother who had dropped out of high school and gotten into trouble with the law, Julie was left back in second grade and took it for granted that she was, in a word, stupid.

"I was never really good at reading and writing," she told me. "Or at anything."

Then, in 1991, Julie entered fourth grade and found herself in the class of a new teacher, Kevin Cripe, who had the outlandish idea that his students were capable of great things.

"When I talked to older teachers," Cripe told me, "they said that Julie was just not very smart. One of her older brothers was in and out of jail. She had been left back. Her younger brother, Danny, had also been left back. And she was not a great reader."

But Cripe had been a lifelong chess player, and when he decided to start a chess club, he invited Julie to participate.

"I had no idea what it was," she said. "I called it 'chest.' I had never heard of it, but I said sure."

Cripe kept their training fun, but challenging, and Julie picked it up with a speed that surprised even Cripe. She began spending hours leaning over a chessboard, lost in thought, thinking not just two or three moves ahead, but ten or more. After two years of practice, when Julie was in sixth grade, Cripe decided that she and two other kids were good enough to enter a local tournament in Bakersfield.

"Here's what I felt as we were going to that first tournament," Cripe said. "There was this other kid named Jordy. A great kid. Both his parents were psychologists. Jordy was a prodigy. He had gone to private elementary schools and played the piano in concerts. His parents had done all the right things. I thought, here's Jordy, he has all this stuff, he speaks French, and here's Julie. Cognitively, I have to think that her brain has never been fully activated or whatever you want to call it. Sort of like a kid who's never really run, never been pushed to do something athletic. I thought, what would happen if we just treat her brain as if it's going to be like his at some point? So I just decided to treat all the kids in the chess club like they're going to be as smart as all the other kids in these tournaments, the ones from the elite private schools. If I didn't believe that, then it's all hopelessness, right? You might as well burn up all the books."

After the students did well at the Bakersfield tournament and at a number of others in California, Cripe decided he would take Julie and the rest of his team to a national chess championship held in Charlotte, North Carolina.

"Don't do this," a fellow teacher begged him. "You will only embarrass these children."

But Cripe took them, and out of eighty teams, his scored in the top fifteen. Among the hundreds of students participating, Julie ended up among the tournament's top ten.

"I didn't start winning till I was thirteen or fourteen," she said. "When I was fourteen, I won a lot of money playing in the tournaments. That's how I bought my first car." Eventually, in her age group, Julie was ranked among the top fifty female players in the United States.

Then her younger brother, Danny, joined the team and soon became its best player. At a national championship held in Tucson, Danny reached the last round, his team clinging to the hope of scoring in the top ten, when the stress got to him.

"He throws up before the last round because he's nervous," Cripe said. "He was the leader. I said, 'Okay, Danny, if you are truly sick, I'll call your mom; we'll withdraw you from the tournament. But if you're nervous, here's what I want you to think about. You have earned this. Everybody else is as nervous as you are. And I want you to enjoy this moment, because there are seven hundred other people here today who have no chance to win a trophy. So what do you want me to do?' And he said, 'I want to try to play.' Then I gave him one last piece of advice: 'If you throw up again, aim for the floor, because if you hit the board, it's going to be hard to play with the chess pieces.'

"He won his game fairly quickly. Every single other student on our team who came out after him also won. They watched Danny win after he threw up. It almost makes me cry every time I talk about it. He was one of the 'dumb ones,' and he finished in the top ten of the national chess championship that year. And our team finished in fifth place. We were ahead of Hunter College Elementary School that year. That's the school in New York City that's always among the best. They were in sixth or seventh place."

Danny went on to graduate from the University of the Pacific with a degree in mechanical engineering. He now works as an engineer for an international manufacturing firm. Julie graduated from the University of Mississippi and is now a homemaker living with her husband, Calbemar, and a young daughter, Isabel.

"I definitely think chess improved my thinking abilities," Julie told me. "And it definitely improved the thinking abilities of other kids in the chess club. We all got better in our grades and everything else. It just had to do with how hard you worked. You get pretty good at it. You sit there for so long. You've got to picture the moves in your head. At the beginning, you can't really think that far into it. When I was really practicing, I could think fifteen, even twenty moves ahead. You have to sit there for hours and try to think through all these different scenarios. And you're just thinking of different consequences. You take that and put it into your own life. If I do this, then this can happen. If I do that, then that can happen. And then you just make the best decision from there."

What, really, is the meaning of intelligence anyway?

"There are some really ignorant people out there," Julie told me, "the people who are prejudiced and think that just because some kids are from a poor area, and their parents didn't have an education, they automatically have to be stupid. And we're not stupid. I'm not stupid. There are lots of smart kids out there. There's lots of things we could get into. It just has to do with the choices you make. That's why I said chess definitely helped me make the right choices."

On the other side of the country, among the most affluent of New York City's parents, another approach to increasing intelligence is being pursued by those able to pay a couple hundred dollars per hour. Founded in 2009, Bright Kids NYC now has as many as five hundred children enrolled at any time, most of them four-year-olds seeking to gain admission to the public schools' gifted and talented program. Although admission was once decided by each individual school district in the city, leading some to question its fairness, in 2008 a uniform, citywide standard was created, based on standardized test scores. (Yes, there are standardized tests for preschoolers.) To gain admission to a neighborhood gifted and talented program, children would have to score in the 90th percentile on the tests. To gain access to the highly sought-after citywide program, with space for just four hundred students in five schools, they would have to score in the 99th percentile. The explicit goal of the new program was to increase the number of accepted children coming from less affluent areas, but it had the opposite effect: more kids overall, and more rich ones than ever, were accepted. So the New York Board of Education tried another fix. In 2013, a new test was added: the Naglieri Nonverbal Ability Test, designed to assess cognitive ability independent of cultural background. The result: even more kids overall, and more rich kids in particular, passed the test. What could be causing the disparity? Although Bright Kids NYC was not the only new tutoring program aiming to help children score well on the tests, it was certainly the largest and most sophisticated, and it had truly stunning results: 94 percent of the children who prepped with Bright Kids scored in the 90th percentile on the tests, and 49 percent of them—nearly half—scored in the 99th percentile. The results suggest a real-life Lake Wobegon, the fictional hometown of Garrison Keillor on his long-running radio show, where "all the children are above average."

As recently as 2008, the consensus among mainstream intelligence researchers was that human intelligence is just too complex, and too closely linked to innate characteristics of the brain, to be significantly modified by any straightforward training method. Sure, they agreed that exposing children to an enriched environment does generally improve their chances for reaching their potential. But not by much. Because unlike a test of physical strength, which measures only how you performed *today*, intelligence tests have always been pitched as an upper limit on what you can *ever* do: a cognitive glass ceiling, a number tattooed on the soul.

And that's why most of us have come to think of intelligence researchers as a bunch of jerks and the IQ test as just plain un-American. Because who wants to be told that we can work and sweat all we want, we can train to run a marathon or learn a new language, we can set a goal and achieve it—but intelligence is the one mountain we can never climb? Then again, perhaps the belief that intellectual disability is heritable and beyond remediation is just the other, darker side of the American spirit: it was in the United States, after all, that the pseudoscience of eugenics had its birthplace, where some sixty thousand sterilizations were performed in the twentieth century, continuing into the 1960s, most of them forced, many of them involving people deemed to be "imbeciles" or "feeble-minded." Championed by the likes of Margaret Sanger, J. H. Kellogg, and Alexander Graham Bell, sanctioned for a time by the U.S. Supreme Court, and funded by such august bodies as the Carnegie Institution and the Rockefeller Foundation, the eugenics movement in this country was credited by Nazi leaders, including Adolf Hitler himself, as inspiring their "war on the weak." Yet even to this day, there remain academics who continue to harp away on the supposed intellectual superiority of one racial or ethnic group over another. As recently as 2009, a dissertation for a Harvard doctorate in public policy asserted: "Immigrants living in the U.S. today do not have the same level of cognitive ability as natives. No one knows whether Hispanics will ever reach I.Q. parity with whites, but the prediction that new Hispanic immigrants will have low-I.Q. children and grandchildren is difficult to argue against." Four years later, the writer of that dissertation, Jason Richwine, authored a study for the Heritage Foundation, a conservative think tank, that criticized immigration reform.

Given all the above, it is not surprising that the general public's view of IQ has pretty well gone down the toilet. In business-speak: intelligence has a brand problem. Popular culture these days has consigned it to the same dark corner into which it has cast pesticides, bullying, and Lindsay Lohan. I caught a whiff of the ill wind blowing against IQ these days in an e-mail from my brother Dave in Maine, who's been ribbing me ever since he heard about the subject of this book:

> Mister Schmarty: Dan, just promise if you get any schmarter, you won't turn into an evil super bad guy like Lex Luthor. Hey, can you add making people nicer to schmarter? James Holmes, schmart, not very nice, the same for Ted Kaczynski. Mister Rogers: very nice, how schmart who knows but wouldn't you like him as your neighbor?

He raises a serious point: a populist vein of American culture has long equated "genius" with "evil" and celebrated a lack of learning as evidence of honesty and decency. These days, even the intelligentsia disdain intelligence, none more so than the writers Daniel Goleman, Malcolm Gladwell, and Paul Tough. In 1995, Goleman published his groundbreaking and hugely influential bestseller, *Emotional Intelligence*, arguing that the ability to "rein in emotional impulse; to read another's innermost feelings; to handle relationship smoothly" is as important as, or more important than, intellectual capacity. Then in 2008 Gladwell published *Outliers: The Story of Success*, in which he made famous psychologist K. Anders Ericsson's research showing that talent plays virtually no role in accomplishment, and that what matters—all that matters—is hard work, specifically ten thousand hours of practice in one's given field. Most recently, in 2012, Tough came out with *How Children Succeed: Grit, Curiosity, and the Hidden*

Power of Character, based on research by psychologist Angela Duckworth and others examining the powerful role of characteristics like self-control, conscientiousness, and determination.

Wonderful insights, all. Hard work, grit, and emotional poise are definitely important to success in life. Nobody can argue with that. But wait a minute: does the importance of those qualities mean that intelligence has no value at all? Certainly IQ is not everything; perhaps it's not even the most important thing, but it's definitely one of them. As we all knew in elementary school and can see in our workplaces and on the front pages of the newspaper every day, intelligence, or smarts, or whatever you want to call it, does matter. Intelligence distinguishes humans from our fellow creatures on Earth. Intelligence—not just knowing a lot of dumb facts, but having the ability to understand and analyze those facts, to learn, to make sense of things, to turn information into knowledge, to turn knowledge into profit, to find meaning in chaos—is power. It's how, tens of thousands of years ago, we mastered fire and learned to farm rather than forage. It's not the only reason, but it's one of the reasons that Warren Buffett, Mark Zuckerberg, and Bill Gates are richer than you are. (Both Zuckerberg, who founded Facebook, and Sergey Brin, who cofounded Google, were selected in adolescence, in part on the basis of scoring high on standardized tests, to attend the Center for Talented Youth at Johns Hopkins, as was Stefani Joanne Angelina Germanotta, better known as Lady Gaga.) It's how Malcolm Gladwell, Daniel Goleman, and Paul Tough wrote such awesome books. Because they're smart, and because, as politically incorrect as it has become in polite society to say so, intelligence still matters.

And not just for school and career achievement. What's surprising, given how we think of intelligence as being all in our heads, is how it contributes to the well-being of our bodies, in ways that are not yet fully understood. A recent study of 1,116,442 Swedish men whose IQs were tested at age eighteen, for instance, found that after twenty-two years, those who scored in the bottom 25 percent were over five times more likely to have died of poisoning, three times more likely to have drowned, and over twice as likely to have been killed in a traffic accident as those who scored in the top 25 percent. Overall, by middle age, for every 15 points lower on the IQ scale that a man's intelligence was at age eighteen, his risk of dying by middle age increased by one-third and his risk of being hospitalized for some kind of assault increased by one-half. In another study, of Scottish adults born in 1921, even after adjusting for the effects of social class and childhood deprivation, every 15-point drop in IQ measured at age eleven was associated with a 36 percent increased risk of death by age sixty-five. In a host of other studies, intelligence has been repeatedly linked to the risk of getting murdered, developing high blood pressure, having a stroke or heart attack—even to early menopause, with one study finding that every 15-point gain was associated with a 20 percent reduction in the likelihood of entering menopause by age forty-nine.

Anyone convinced that intelligence doesn't matter should try telling that to the 800,000 children and adults in the United States receiving Social Security income due to a diagnosed intellectual disability.

Try telling the 250,000 service members diagnosed with a traumatic brain injury since 2000 that intelligence doesn't matter. And I don't mean the kind of pointy-headed academic test-taking ability that the very word "intelligence" connotes, but the mental sharpness and insight that those tests measure, and which are the very ones impaired by a brain injury.

Try telling the 5 million Americans who are losing not only their long-term memory but their ability to follow a conversation and balance their checkbook due to Alzheimer's that intelligence doesn't matter. (By the way, the smarter you are, the later in age you are ever likely to be diagnosed with Alzheimer's, due to something researchers call "cognitive reserve.")

Try telling people with major depression or schizophrenia that intelligence doesn't matter. One of the most incapacitating aspects of their diseases, surprisingly enough, is the significant intellectual impairments they cause, so much so that those with the strongest remaining cognitive abilities generally have the best prognoses for recovery.

All of which would be thoroughly depressing and discouraging if we could do nothing about our intelligence, as we have been so long told. Given the supposedly unyielding nature of this albatross called intelligence, it's no wonder we have, as a culture, decided to do our best to ignore it, as we do death.

But what if all the experts who have told us for a hundred years that it cannot be changed are wrong? What if the brain is like pretty much every other part of the physical world, in that human ingenuity can find a way to tinker with it? Think about it: we can transplant a heart, construct a bionic retina to let the blind see, and build robotic legs to permit the lame to walk; we can get breast implants and have a sex change. But we can't increase our brain's functional abilities? Are smartphones the only thing we can make smarter? What is this intelligence thing anyway: is it some kind of forbidden fruit from the Tree of Knowledge? Does it not have a real, physical basis? Are these researchers who tell us it can never be changed actually scientists—or are they high priests of an IQ cult?

Are we not smart enough to figure out how to make ourselves smarter?

The first new answer in a century to that question came in May 2008. Two Swiss researchers by the names of Susanne Jaeggi and Martin Buschkuehl published a study that month in the prominent

Proceedings of the National Academy of Sciences reporting on what happened when college students played a peculiar computerized game called the N-back for twenty minutes a day, five days a week, for four weeks. The game—about which I'll go into greater detail in the first chapter—was designed as a test of something called *working* memory: a person's moment-by-moment attention and the ability not simply to remember short-term but to juggle and update and manipulate and analyze the content of those memories; that is, to *work* with them. In Jaeggi and Buschkuehl's study, this test of working memory was turned into a tool for training, and sure enough, the longer the students practiced the N-back game, the better they got at it. Importantly, however, before and after those four weeks of practice, the students took a test of a mental ability called *fluid* intelligence. Standard IQ tests include measurements of *crystallized* intelligence, your treasure trove of stored-up information and how-to knowledge, which just keeps growing as you age—the sort of thing tested on *Jeopardy!* or put to use when you ride a bicycle. Fluid intelligence, on the other hand, is the underlying ability to learn, the capacity to solve novel problems, see underlying patterns, and figure out things that were never explicitly taught. It has long been known to peak in early adulthood, around college age, and then gradually decline (which is why the most influential work of mathematicians, physicists, and musicians usually occurs in their twenties and then quickly falls off). And unlike physical conditioning, which can transform ninety-eight-pound weaklings into hunks, a hundred years of scientific doctrine insisted that fluid intelligence was impervious to the effects of training. Yet in Jaeggi and Buschkuehl's study, after just four weeks of doing the N-back, the students' scores on a measure of fluid intelligence increased, on average, by 40 percent.

"Increasing Fluid Intelligence Is Possible After All," stated the headline of an editorial accompanying the study, which drew wide coverage in the media and sparked the academic equivalent of a food fight among intelligence researchers. Derided and ridiculed by old-school researchers as the equivalent of "cold fusion," it also drew strong praise from many younger ones. Like controlled flight prior to the Wright brothers, the notion that human intelligence could be increased struck some as laughable, others as inevitable.

In the years following publication of Jaeggi and Buschkuehl's findings, a grand total of four randomized, placebo-controlled studies have been published (as of this writing) finding no benefit of cognitive training. Skeptics point to these four studies as evidence that training remains a fool's errand. Yet in contrast, by my count, seventy-five other randomized, placebo-controlled studies have now been published in peer-reviewed scientific journals confirming that cognitive training substantially improves intellectual abilities. Twenty-two of those studies specifically found improvements in fluid intelligence or reasoning, while the remaining fifty-three found a variety of other significant benefits in abilities such as attention, executive function, working memory, and reading. Results have now been seen not only in elementary-school children, but in preschoolers, college students, the middle-aged, and the elderly. Healthy volunteers have benefited, as have people with disorders including Down syndrome, schizophrenia, traumatic brain injury, alcohol abuse, Parkinson's disease, chemotherapy-treated cancer, attention-deficit/hyperactivity disorder (ADHD), and mild cognitive impairment (a common forerunner of Alzheimer's disease). Gains have been seen to persist for up to eight months after the completion of training.

Even for those concerned about emotional intelligence, short-term cognitive training has been shown to pay off. In March 2013, the *Journal of Neuroscience* published a randomized study by Cambridge University researchers showing that people who spent just twenty days training for about a half hour a day on a version of N-back that incorporated emotion-laden words like "dead" and "evil," as well as images of faces displaying fear, anger, sadness, or disgust, significantly improved their performance on a gold-standard measure of emotional control, called the emotional Stroop task. Those gains, by the way, were accompanied by greater activity in the part of the frontal lobes associated with emotional regulation, as revealed by fMRI brain scans.

Despite the lopsided evidence in favor of training's effectiveness, the dispute among scientists over whether the gains are real remains fierce, at times downright ugly. As a science journalist, I have been privileged to be present during some of the most, shall we say, piquant debates, and to speak with most of the leading voices in the field, on both sides of the divide. I've now interviewed a couple hundred researchers in the United States, Britain, France, Germany, Japan, and China. I visited Walter Reed National Military Medical Center, where I met brain-injured veterans. I went to the San Francisco offices of Lumosity, the biggest online provider of these cognitive games aimed at improving intelligence. And I met twice with the guy who leads the funding in this area at the Intelligence Advanced Research Projects Activity, or IARPA. It's a government intelligence agency, like DARPA for spies. The guy who is funding research on this is hoping they can figure out how to make their intelligence officers more intelligent, so they can see the danger in Benghazi before the chief diplomat is killed.

But he has a problem. The field is in such an uproar that each time I met him, the IARPA guy asked me what I think is really going on. Essentially what he was asking me was: does this stuff really work? And I'm going to tell you what I told him: before I put my name on a book saying that something as basic to a person's nature as intelligence can actually be improved in a matter of weeks or months, the skeptical bastard in me demands that I personally test these methods on myself first. Which I did, and will report on, for better or worse.

This book, however, is not about me: it's about the field of intelligence research as it undergoes a revolution, as an ever-growing majority of researchers shifts from viewing fluid intelligence as something unchangeable, like eye color, to something more like muscular strength, which has a biological basis but is equally susceptible to training. It's a startling transformation in our understanding of a fundamental human trait: the capacity for rational thought—the ability to learn—and whether a strict limit is set for each of us on the day of our birth, or whether we can do something about it. The overturning of the pernicious dogma that our intelligence is unchangeable holds enormous implications for every level of society: young and old, rich and poor, genius and cognitively disabled alike. No one is saying that cognitive training can turn an intellectually disabled person into a genius. Exactly how much people can benefit, and which methods work best, remain a work in progress But that shouldn't be surprising. Dated to Jaeggi and Buschkuehl's 2008 study, the new science of building brain power is barely six years old. This book tells the story of the birth of that science and what it may mean for anyone who ever wanted to be smarter.

CHAPTER 1

Expanding the Mind's Workspace

Our story begins in June 1997 on a kayak floating on Mälaren, Sweden's third-largest lake, its nooks and crannies sprawling for more than fifty miles west of Stockholm. Paddling the kayak was Torkel Klingberg, a graduate student in the department of psychology at Sweden's most prominent research facility, the Karolinska Institute, who had just completed a study pinpointing where in the brain problems requiring working memory are solved. Then, as now, research psychologists and neuroscientists were engaged in trying to do for the brain what pioneering anatomists had done hundreds of years earlier for the rest of the body: figure out which parts do what. Using a kind of imaging technology known as positron-emission tomography to see inside the brain, Klingberg found that no matter which kind of working-memory task he gave volunteers, or even whether the information was presented by sound or sight, the same six brain regions kept showing increased blood flow—that is, increased workload—most of them in the frontal lobes, located behind the forehead.

Taking a day off after completing this study, Klingberg went paddling along in the nearly perpetual daylight of the Scandinavian midsummer. As he paddled, a question buzzed around his head: what does it mean that the same areas of the brain are engaged in all these working-memory tasks? Questions such as these, big as Mälaren itself, are the kind that lesser scientists avoid, lest they get lost in idle speculation. But Klingberg, who could pass for Lance Henriksen, the actor who played the android Bishop in the film *Aliens*, kept puzzling over the question until an answer came to him—not so much an answer, really, as a hypothesis. If the same areas of the brain are involved in all working-memory tasks, he reasoned, then perhaps training on one such task should result in improvement on another, because both require strengthening of the same brain region. Just as doing push-ups makes a person better at lifting weights.

Klingberg made a note of this hypothesis in a small black book that he carried around with him. It sat there for two years until Klingberg joined the Karolinska's department of neuropediatrics, in 1999, to pursue his doctorate. Because the department conducted many studies of ADHD, Klingberg now had access to volunteers with whom he could test his idea.

But he had a problem: other psychologists had already supposedly proved that Klingberg's experiment would never work—that practice on one short-term memory task never transfers to improvement on another. Most famously, K. Anders Ericsson and colleagues at Carnegie Mellon University published a study in 1980 in the eminent journal *Science* describing their twenty-month experiment with a young man they identified only as S.F. An undergraduate "with average memory abilities and average intelligence for a college student," S.F. volunteered to see if his short-term memory could be substantially increased. With no instruction on memory strategies, he was asked to listen to a series of random digits and then recite back as many of them as he could. At first, like most people, he could accurately remember only seven. ("The Magical Number Seven, Plus or Minus Two" was the title of a classic 1956 paper by psychologist George A. Miller, in which the limit of items that humans can hold in short-term memory was first described.) But as S.F. practiced for about an hour a day, three days a week, for over a year and a half, he gradually was able to successfully remember more and more numbers. After fifteen weeks, he could accurately recite back up to twenty-five random digits, in sequence. After a year, he could recite back seventy in a row. Eventually, after twenty months, he reached ninety numbers in a row, equal to some of the best memory champions, and his rate of improvement showed no signs of slowing. Yet when he tried to remember anything other than a string of random numbers, even a string of letters, he was no better than anyone else: "His memory span dropped back to about six consonants."

How could that be? The key to understanding how S.F. could learn to remember ninety digits, but still only six letters, was that he had spontaneously developed mnemonic strategies to turn the random string of numbers into larger chunks that he could remember as either running times, ages, or dates. But of course that strategy, specific to numbers, was of no help when he tried to remember letters or anything else. These kinds of memory tricks, which were also used by the journalist Joshua Foer to win the 2006 U.S.A. Memory Championship, as described in his bestselling book, *Moonwalking with Einstein*, are powerful as far as they go. But in the end, they are tricks. They help you remember lists of stuff. But they do not help you make sense of those lists. They do not make a person smarter. They do not improve working memory.

Here, perhaps, I should make clear the important distinction between short-term memory and working memory. It's a distinction that many journalists writing for a general audience, and even some psychologists, often fail to note. Both occur over a matter of seconds, certainly not hours, let alone months or years. Short-term memory is what Ericsson was measuring: the ability to quickly spit back some stuff that you have been given. It's simple. And, surprisingly, it has very little to do with intelligence and problem solving. Working memory, on the other hand, is your ability to *manipulate* the stuff you're being asked to remember: flipping the numbers around, adding them up, deciding whether they're odd or even. With language, working memory enables you not simply to remember this sentence, but to understand its meaning and consider its implications. As one researcher has called it, working memory is the mind's workspace, the factory floor where raw materials get processed and assembled into something useful. Short-term memory enables you to remember a telephone number, but working memory empowers you to multiply the first three digits of that number by the last four digits in your head. Doing so, critically, requires exquisitely tight control over the objects of one's attention and the ability to avoid distraction. The demands of working memory explain why doing multiplication of two-digit numbers in your head (let alone four-digit numbers) is so difficult: because you have to do it in parts, and set aside the solution to one step while solving the next, placing things into the back of your mind, out of your conscious awareness, and then rapidly pulling them back to attention as necessary. Working memory is what permits a poet to play with words to discover the best expression of a given thought; it's how we remember the second and third steps of a set of directions after completing the first. The limits of working memory explain why driving a car while using a hands-free cell phone is just as dangerous as holding the phone in your hands: because your ability to make sense of things is a precious, limited commodity.

The most colorful and outrageous example I have ever seen of a powerful working memory in practice is credited to my oldest friend, Dan Feigelson. Beginning when we were teenagers, he discovered that he could, on demand, say any word backward, no matter how many syllables. You could say "incompatibilities" and, a few seconds later, he would say, "seitilibitapmocni." It's astonishing and hilarious to witness, and his secret, he told me, is that he visualizes the word as if it's written on a chalkboard and then simply reads it backward.

That's working memory.

What Ericsson had concluded from his study of S.F. was that training does not ultimately increase the general capacity of short-term memory. But what Klingberg wanted to know was whether something other than strategies and tricks could be used to increase the general capacity for working memory.

On that question, he drew inspiration from one of the most influential and honored figures in the history of research into neural plasticity: Michael Merzenich. In the early 1980s, when most neuroscientists still believed that virtually all areas of the brain were permanently hard-wired to handle only particular types of information, Merzenich published studies showing that, in a matter of weeks, he could change which areas of a monkey's brain handled information from, say, the first digit of its left hand—simply by disabling the second digit. Rather than sitting idle when nerve signals stop coming, the area of the brain previously devoted to one finger begins processing information from another. Over the following three decades, Merzenich built on this observation to show that animals, including humans, could benefit from neural reassignment: as more attention is given to distinguishing between pinpoint differences in touch, sound, or sight, the area of the brain devoted to that distinction expands and, in the process, gets better at it. Dyslexic children, he found, could be trained to discern subtle differences in sounds to enable them to better understand spoken language; elderly drivers in their seventies could likewise be trained to regain the wider field of view that they had gradually lost over a period of decades.

From Merzenich's groundbreaking research, Klingberg took two principles. First, to be successful, training should be offered in relatively short bursts of twenty to thirty minutes a day, but repeated four to six times a week for at least four weeks. Second, the training schedule should be continuously adapted to the capacity limit of the individual being trained. It can't be too easy; it can't be too hard; it has to be right at the edge, and it has to *stay* at that edge, getting harder as the person gets better. Together, these two principles developed by Merzenich made for a standardized regimen: four weeks of short daily bursts of intense training that is continuously adapted to remain always at a person's capacity limit. That regimen would prove crucial not only to Klingberg's progress, but to the entire field's.

For the purposes of his study, aimed at training working memory, Klingberg enrolled fourteen children between the ages of seven and fifteen diagnosed with ADHD by a pediatrician. All of the children were asked to spend twenty-five minutes per day, five days a week, for five weeks playing a variety of computerized working-memory games designed from scratch by a programmer, Jonas Beckeman. But half of the kids played games that adaptively got harder to remain at their capacity limits, while for the other half, the games started easy and remained easy. Each of the games was a variant of previously standardized tests of working memory. With the "backwards digit span," for instance, a list of numbers is shown on a keyboard and simultaneously read aloud, and the child has to type back the digits, but in reverse order. (That's what makes it a working-memory task rather than a simple measure of short-term memory, because the list of numbers has to be mentally manipulated and recited in backward order.) For the adaptive training group, the list of numbers kept getting longer as the children grew better at reciting them backward.

To old-school psychologists, the experiment sounded like an exercise in futility. Tasks like these had been developed to serve as the mental equivalent of vision tests, not as training programs. Practicing them made about as much sense as practicing an IQ test over and over, since improvement on the test would not mean you were actually getting smarter, but only that you were getting better at taking the test.

But here is where the results proved startling: the seven kids who trained adaptively not only got better on the trained tasks, they also improved on other measures of working memory. It was as if they practiced golf and got better at basketball. What's more, their hyperactivity, as measured by head movement, was also significantly reduced. (Other studies have found that children diagnosed with ADHD generally perform worse on tests of working memory than other children do, but the two are not synonymous: roughly as many girls as boys have a low working memory, but far more boys than girls are diagnosed with ADHD.) Most incredibly, even bizarrely by the standards of orthodoxy then holding sway, the trained kids in Klingberg's study also did much better on the Raven's progressive matrices, long regarded as psychology's single best measure of fluid intelligence. If the results were to be believed, the kids had gotten smarter.

"This is impossible. This doesn't work."

In June 2002, having just completed the Swiss equivalent of a master's degree in psychology at the University of Bern, Switzerland, Martin Buschkuehl was searching for a topic for his PhD dissertation when he came across a study whose very title seemed a contradiction in terms. Tall, blond, and good-looking—in other words, a typical Swiss—Buschkuehl had grown up rowing competitively in Lucerne. During high school, he won a Swiss national championship three years in a row, and also won twice as part of the Swiss National Rowing Team at the French championships. Having trained for years to reach and exceed his physical limits, his studies in psychology naturally gravitated in that direction. But, he knew, there are some limits that cannot be exceeded, because they are traits, defining characteristics of an individual, which are not subject to change. Blue eyes cannot be trained to turn brown. Men cannot be trained to turn into women. And working memory—the hard, immutable kernel at the center of fluid intelligence—cannot be trained to increase. Yet here was a study in the *Journal of Clinical and Experimental Neuropsychology*, by some guy named Torkel Klingberg, claiming to have done just that: "Training of Working Memory in Children with ADHD." After five weeks, twenty-five minutes a day, of practicing a bunch of goofy little working-memory tests, the kids were smarter and less hyperactive?

"This is impossible," Buschkuehl muttered to himself after reading the paper. "This doesn't work."

He showed the paper to his girlfriend and fellow psychology graduate student, Susanne Jaeggi. Given to plaid shirts, corduroy pants, and the sort of sturdy shoes fit for hiking in the Alps, Jaeggi could be a poster child for intellectuals, wearing no makeup and little jewelry, her long, straight brown hair parted in the middle and hanging down past black-framed, functional glasses.

"I don't believe it either," she told him. "That's weird."

Yet they were both intrigued. After all: what if it were true? If training on a working-memory task could transfer to an increase in fluid intelligence, it would be cognitive psychology's equivalent to discovering particles traveling faster than light: all but unbelievable, but hugely important.

And really, the weird little study seemed tailor-made for Buschkuehl and Jaeggi to follow up. He was already involved in a study of improving octogenarians' well-being; training was his thing. And working memory was Jaeggi's area of interest; she was conducting various studies of people's abilities using her own favorite working-memory test, the N-back. Perhaps, they agreed, they should see if they could run a study with the N-back as their training task.

This N-back is quite the beast, not only to do but to describe. Ten seconds of trying it for yourself on one of the many versions available online will help you understand it far better than spending ten minutes reading about it. But here goes: Imagine that you are listening to a string of letters spoken aloud. You are asked to press a button every time you hear the same letter repeated twice in a row. That's 1-back. That's easy. So if you hear the list *n-a-m-m-a-m*, you press the button when you hear the second *m*, right? But now let's try 2-back: this time, you have to press the button when you hear the *last* letter in the series, because this last *m* was preceded two letters earlier (hence "2-back") by another *m*. If you were being tested on 3-back, however, you would press the button when you heard the second *a*, because it was preceded three letters earlier by the first *a*. And so it goes, to 4-back, 5-back, and on.

What makes this task so difficult is that the list just keeps coming at you—not a short list of six letters, like the one I have presented as an example, but a list that continues, with another letter and another letter, for a minute and a half. So you are constantly updating and keeping track of the current sequence of two, three, four, or more letters, which is constantly changing as the next one is added. It requires total concentration. Let your mind drift for a moment and you're lost.

But wait. In order to make it devilishly harder, Jaeggi and Buschkuehl decided to use what's called the *dual* N-back task. As you hear this random sequence of letters, you also see a dot on your computer screen moving randomly among eight possible spots on the outer squares of a tic-tac-toe board. Now your mission is to keep track of both the letters *and* the dots as they just keep coming. So, for example, at the 3-back level, you would press one button on the keyboard if you recall that a spoken letter is the same one as was spoken three times ago, while simultaneously having to press another key if the dot on the screen is in the same place as it was three times ago.

That's right. Ouch.

The point of making the task so difficult was to literally boggle the mind, to overwhelm the usual task-specific mental strategies that people develop for math, crossword puzzles, Scrabble, and the like. If people got better as they practiced the dual N-back, they figured, perhaps they would be actually expanding their working memory.

Just as Klingberg had borrowed from Merzenich, so Buschkuehl and Jaeggi borrowed from Klingberg the regimen of having the participants do their dual N-back training task for about twenty-five minutes per day, five days a week. Likewise, the computer program that Buschkuehl devised always kept the N-back level adapted to each participant's capacity. If the person could accurately keep track of both the spoken letters and the dots on the tic-tac-toe board at the 2-back level, he or she would automatically be moved up to 3-back, and so on.

After enrolling a couple of dozen undergraduates from the University of Bern, they first tested their volunteers' fluid intelligence with the Raven's progressive matrices. Anyone who has taken an intelligence test has seen matrices like those used in the Raven's. Picture three rows, with three graphic items on each row, made up of squares, circles, dots and other symbols. Do the squares get larger as they move from left to right? Do the circles inside the squares become filled in, from white to gray to black, as they go downward? One of the nine items is missing from the matrix, and your task is to discern the underlying patterns—up, down, across—in order to select the correct item from one of six possible choices. While at first the solutions are obvious to most people, they get progressively harder, reaching the point where, by the end of the test, they baffle all but the brainiest.

Why matrices should be considered the gold standard of fluid-intelligence tests may not be obvious at first. But consider how central pattern recognition is to success in life. If you're going to find buried treasure in baseball statistics, permitting your team to win games by hiring players unappreciated by other teams, you'd better be good at matrices. If you want to find cycles in the stock market to exploit for profit; if you want to find the underlying judicial reasoning behind ten cases you're studying for law school—for that matter, if you need to suss out a woolly mammoth's nature in order to trap, kill, and eat it—you're essentially using the same cognitive skills tested by matrices.

After the undergraduates took the Raven's, they each agreed to drop by the psychology department's testing laboratory for a half hour per day, five days a week, for four weeks, in order to train on the N-back. Within days, most of them had jumped from mastering the 3-back to dabbling with the 5-back. By the end of the four weeks, some got as far as 8-back. And afterward, when they took the Raven's again, their average scores had jumped by over 40 percent.

Skeptical of even their own results, yet impressed by how easily they had achieved the seemingly impossible, Jaeggi and Buschkuehl wrote up their dissertations, obtained their doctorates, and accepted an invitation to pursue their postdoctoral research at the University of Michigan, in the laboratory of John Jonides, professor of psychology and neuroscience. There they repeated the dual N-back experiment, this time adding a placebo control group whose fluid intelligence was twice tested with progressive matrices, without undergoing the training. They also decided to see if they could detect the kind of dose effect that is routinely measured in drug studies, so that the more training people did, the higher their resulting fluid intelligence. Sure enough, at the conclusion of the study, those who practiced the dual N-back for just twelve days saw on average a gain of a little over 10 percent on their matrices test. Those who practiced for seventeen days gained more than 30 percent, and those who practiced for nineteen days increased an astonishing 44 percent.

They finally published the results of their study in the *Proceedings of the National Academy of Sciences*, on May 13, 2008. Unlike Klingberg's study, which had received little notice by the popular press, Jaeggi and Buschkuehl's study became an immediate sensation, making headlines in newspapers around the world. "'Brain Training' Games Do Work, Study Finds," announced the British newspaper the *Daily Telegraph*. "Memory Training Shown to Turn Up Brainpower" was the headline in the *New York Times*. The attention came for a number of reasons, including its bold title ("Improving Fluid Intelligence with Training on Working Memory"), the prominence of the journal, the elegance of Jaeggi's writing, the statistical rigor of the study, and the accompanying celebratory commentary by Robert J. Sternberg, then dean of the Schools of Arts and Sciences at Tufts University and himself a highly regarded intelligence researcher. "Jaeggi et al. have made an important contribution to the literature," Sternberg wrote, "by showing that fluid intelligence is trainable to a significant and meaningful degree; the training is subject to dosage effects, with more training leading to greater gains; (and) the effect occurs across the spectrum of abilities, although it is larger toward the lower end of the spectrum. Their study therefore seems, in some measure, to resolve the debate over whether fluid intelligence is, in at least some meaningful measure, trainable."

•—•

Somehow I had missed all the excitement. It wasn't until three and a half years later, in 2011, after I had written an article about drugs being tested to increase intelligence in people with Down syndrome (about which, much more to come in chapter 9), that I became interested in the possibility of increasing intelligence in those of us without a diagnosed intellectual disability. By then, Jaeggi and Buschkuehl's study had, quite simply, revolutionized the field of intelligence research, with hundreds of subsequent studies citing it.

"My findings support what they've done," Jason Chein, assistant professor of psychology at Temple University in Philadelphia, told me when I reached him by telephone. Chein had seen improvements in cognitive abilities after training people not with N-back but with other working-memory tasks, the verbal and spatial complex-span tasks. "I've never replicated exactly what they do. But across a number of labs, using similar but different approaches to training, we have had related successes. Cautious optimism is the best way to characterize the field now."

Even the U.S. military had jumped in to see if the cognitive abilities of officers and enlisted men and women could be increased. Harold Hawkins, a cognitive psychologist at the Office of Naval Research, was in charge of funding research in the area and had already approved grants to Jaeggi and a half dozen others. "Up until about four or five years ago, we believed that fluid intelligence is immutable in adulthood," Hawkins told me. "No one believed that training could possibly achieve dramatic improvements in this very fundamental cognitive ability. Then Jaeggi's work came along. That's when I started to move my funding from some other areas into this area. I personally believe, and if I didn't believe it I wouldn't be making an investment of the taxpayers' money, that there's something here. It's potentially of extremely profound importance if it is there."

With Jaeggi and Buschkuehl already having replicated their findings in studies of elementary-school children and older adults, and with commercial enterprises sprouting up to offer cognitive training online, in tutoring centers, and through trained psychologists, I decided to call and interview Jaeggi by telephone. I asked if I could come out to meet them, and she agreed. Then I asked if she would be willing to help me conduct my own journalistic test of the N-back and other methods shown to benefit cognitive abilities. Would she be willing to test my fluid intelligence before I began my training regimen, and again afterward?

"You should know first that some people have a really hard time training on N-back," she warned me. "They say it's really frustrating and really challenging and tiring. They really have a hard time to stick with the training."

"What's been your own experience training on N-back?" I asked.

"Oh, I haven't," she said. "I've practiced on the task just to learn it, but not really to train. I'm fine with how smart I am already. And anyway, we've tended to see the biggest effects for those at the lower ability levels, so long as they put in the effort. So you are going to have a tough time showing much effect."

Her lack of interest in training herself surprised me, but it turned out to be universal among researchers involved in the field: neither Jason Chein, John Jonides, nor any others I met ever confessed to training. Some echoed Jaeggi's sentiment about the greatest benefits being seen for the least intelligent people. But I knew that many of their studies contradicted that claim, since they often involved graduate students at prestigious universities. Were they just too proud to be seen as needing or wanting to increase their own intelligence?

But if I was willing to try, Jaeggi said, she was willing to conduct before-and-after tests of my fluid intelligence and to provide me with their version of the N-back.

Game on. And so on Halloween of 2011, I flew to Detroit, rented a car, and drove out to Ann Arbor to meet Jaeggi, Buschkuehl, Jonides, and their colleagues.

•—•

"I had this big project to train all the dolts."

Buschkuehl, Jaeggi, and I were talking over lunch when I heard him make that outrageous comment.

"Excuse me?" I asked.

"I was training old adults," Buschkuehl said.

"Oh," I said. "Right."

"Octogenarians," he continued. "Someone else was going to offer them resistance training, strength training, and he asked me if I was interested in trying something else. I was always interested in making people better, in how people can get beyond their capacity limits."

"You were a rowing coach, too," said Jaeggi.

"I just liked to find ways how to optimize performance," he continued. "For instance, being able to memorize things better. Being able to solve problems quicker. Enhancing your general ability to handle things."

They had taken me to their favorite pizzeria in Ann Arbor. They said it was the best pizza they had eaten outside of Naples, where Jaeggi's brother lived. She recommended any of the varieties, except the one with truffles. The very mention of truffles made her wrinkle her nose in disgust.

One of the first critical questions I posed was how to pronounce Jaeggi's name.

"Nobody pronounces it right," she said. "It's *YAH*-kee. The Germans, they would say *YAY*-ghee. There are four languages in Switzerland: German, French, Italian, and Romansch. My parents were from Bern, so I spoke Bernese German. But where I grew up, in a little farming village in the Alps called Ftan, everyone spoke Romansch, so I understand that, too."

What I didn't understand was why they'd decided to jump into the field of cognitive training when so much evidence at the time suggested it was impossible.

"I think it's just such an interesting issue," Buschkuehl said, "to train and improve our capacity limits. There are so many questions. How do people react when they get to their capacity limits? What are the neural correlates when you reach your capacity limits? So when we read this report from Torkel Klingberg, it was the first report of its kind. There was nothing else like it out there. I decided to try something like it in the octogenarians."

The working-memory task he devised for them was something Buschkuehl created specially for the seniors; he called it the "animal span task." He designed a computer program to display pictures of different species of animals—donkeys, dogs, cows, ducks—each one shown either upside down or right side up. As each picture flashed on the screen, the participant had to quickly press a button to indicate its proper orientation. Then, after a series of animals had been presented, the participant had to correctly select the order in which the species had been presented.

"The hard part," Buschkuehl said, "is that you have to do two things at once. You have to make a decision about the orientation of the animal. And at the same time, you have to encode which animal follows the next, the sequence."

"Did the octogenarians improve?" I asked.

"They got better," he said. "And we also saw improvement on some similar tasks. There was also a trend toward improved episodic memory. It was not very strong, but for a start this was a nice result."

It was enough that he and Jaeggi decided they could combine her expertise on the N-back with his interest in training to test some undergraduates at the University of Bern.

"Our research interests converged at that point," he said.

"So are two heads better than one?" I asked.

"You know, I think for the university at Bern, and here at the University of Michigan, we're a good deal," Buschkuehl said. "We never stop working."

"We work in the evening," said Jaeggi. "We work on the weekend." Jaeggi's name was listed as the first author on the 2008 paper and had likewise appeared first on subsequent papers involving children and older adults. She is therefore the one most often mentioned by others in the field when describing the study. She insists, however, that Buschkuehl is her equal partner in the research.

"Martin is more of the software developer and the methods guy," she said. "I'm more, I don't know, I write or do something about theory and I do organizational stuff."

Did they never feel competitive or jealous of each other? They both said no.

"I never think about competition," Buschkuehl said. "Life would be too hard."

After lunch, we headed to the office they shared in the windowless basement of the University of Michigan's psychology building. On the door was a cartoon showing a brain with a smiling face and tiny arms and legs. The brain-man was lifting weights over his head. Large letters underneath stated: "Brain Gym."

We were joined by Jonides, the professor of psychology and neuroscience who had invited them to pursue their postdoctoral research in his laboratory and had coauthored their 2008 paper (along with Walter J. Perrig, their academic supervisor in Bern). Trim, with salt-and-pepper hair, Jonides eschewed the sneakers and hoodies that Jaeggi and Buschkuehl were partial to, wearing crisp khakis, brown leather shoes, and a pea-green dress shirt with a little white sailboat emblem. His eyeglasses perched atop his head, Jonides sometimes leaned back against the wall, his arms folded behind his head, and other times hunched forward, gesturing vigorously. A generation older than his two postdocs, he evinced a hard-won wisdom about the ways in which scientific disputes often resemble political ones.

"There are certainly skeptics of the possibility that working memory can be trained in a way that increases fluid intelligence," he said. "There are some who say they can't replicate our results. They say the data shows that intelligence is mostly genetically determined. But we've got a story to tell. All of us have given talks at conferences about this work. When we give talks, although we're careful to air the dirty laundry, the data that don't fit, nonetheless we're still pitching a certain story."

And so he pitched away.

"There are two observations that are worth taking very seriously," he said. "One is that other characteristics are heavily genetically determined. Take height. We know height is 70 or 80 percent genetically determined. Yet we also know there are powerful environmental influences on height. Like nutrition. So even if intelligence is highly heritable, that doesn't mean you wouldn't be able to modify it.

"Another observation is the phenomenon I've called 'dumber over the summer.' If you test kids in April and again in September, they score worse in September. That means that doing nothing, spending the summer watching TV, can certainly influence your intelligence in a negative way. So the story is that you can move intellectual functioning around. You can move it around for the worse, and you can move it around for the better. And who knows why some of them work and some of them don't. I don't doubt that you have to kiss a lot of frogs, but some of them could turn out to be princes.

"Just to give you a completely wild-eyed example, do you know a fellow in Toronto, Glenn Schellenberg? Glenn now has two papers, and I think these are among the best-done studies of their kind, showing that through musical training, he can improve kids' intelligence. Now there's a really unlikely prince. What Glenn is doing is training something that by anyone's estimate should have nothing to do with intelligence, but he's found an effect."

Given the history of failures in efforts to increase intelligence, I asked Jonides why he had decided to get into it.

"A lot of scientists go through the same career cycle I've gone through," he said. "I spent the lion's share of my career studying basic science—basic aspects of mental functioning. Nothing to do with training. Now I'm interested in finding out how I can help the rubber hit the road."

His focus of study for twenty-five years, he said, has been a mental ability, underlying not only intelligence but also many behaviors and emotions, known as cognitive control.

"Right now," he said, "if I were hungry, I'd be thinking about going out to the main lab and sneaking into the kitchen to get some candy. But I'm inhibiting those impulses and continuing to have this conversation. That's an example of cognitive control. That and working memory are at the heart of intellectual functioning. They are part of what differentiates us from other species. They allow us to selectively process information from the environment and to use that information to do all kinds of problem solving. But cognitive control is not just an intellectual matter. In depression, people can't stop thinking these negative thoughts. And the problem with people who can't delay gratification, who become obese or develop an addiction, is that they can't get a thought about some desire out of their mind. All of these are cases where people have lost cognitive control. So I'm involved in studies now that seek to help people regain control."

In Jonides' view, N-back was a method for strengthening a person's cognitive control, the ability to focus attention and avoid distractions. Jaeggi and Buschkuehl shared that perspective.

"We see attention and working memory as like the cardiovascular function of the brain," said Jaeggi. "If you train your attention and working memory, you increase your basic cognitive skills that help you for many different complex tasks."

How long, I asked, do the benefits last?

"We think of it like physical training," Jaeggi said. "If you go running for a month, you increase your fitness. But does it stay like that for the rest of your life? Probably not. You have to keep training."

Does motivation play a role in the effects of training?

"We think so," Jaeggi said. In a study of elementary- and middle-school children published in 2011, they found that only the children who engaged in the training enough to improve significantly on the N-back task saw corresponding gains in their fluid intelligence. "Figuring out how to get more people motivated enough to stick with this kind of training is a challenge. It's a problem, because training doesn't work if you don't do it."

Does the training actually have a physical effect on the brain?

"I'm glad you asked that," Jonides said and grabbed his laptop from the desk. After a few clicks, he turned the screen toward me. "We recently did fMRI scans of people's brains while they were doing N-back. Here's the average activation before a week of training." The image showed sections of a brain lit up in shades of green, yellow, and orange. "And here's what happens after training." He clicked and, even to my untrained eye, the next image showed significantly less orange and more green. "There's a dramatic reduction in activation," Jonides said, "both in the front of the head and the back of the head, suggesting that they're now doing more with less; they've gotten more efficient at doing the N-back task."

He put the computer back on the desk.

"So on that 2008 study," I said, "where the students increased their score on the matrices test by 40 percent, does that mean they literally became 40 percent smarter?"

"I would certainly not say that," answered Jaeggi. "We used just one measure of intelligence or reasoning behavior. What we need to do in the future is to incorporate some real-world measures to really find out about the impact."

"But these matrices are the gold standard to measure fluid intelligence," said Buschkuehl. "And we have many anecdotal reports from our subjects. It's not uncommon for them to say they now understand papers better that they have to read for a class. If people feel like this after four weeks of training for twenty minutes a day, I think that's an impressive effect."

But I would soon see for myself, beginning the next day, when they would measure my fluid intelligence and set me up to road-test the N-back. It had all seemed like a great idea when I discussed it by telephone with Jaeggi, but now that I was here, and the testing was soon to begin, I found myself wondering: what if my IQ is embarrassingly low?

They promised to be ready to test me at 9:00 sharp the next morning. Then, after hours of talking about intelligence, I walked outside into the early darkness of Halloween, past three college students dressed as beer bottles.

CHAPTER 2

Measure of a Man

"How can I observe love?"

Randall W. Engle, one of the most influential living American psychologists, whose research on the relationship between working memory and fluid intelligence set the stage for Klingberg's, Jaeggi's, and Buschkuehl's breakthroughs, sat in the back of a cafeteria at Rutgers University in New Brunswick, New Jersey, where he was scheduled to give a speech nearby, trying to explain one of the most enduring and profound problems in psychological research.

"Most of the things that psychology talks about, you can't observe," he told me. "They're constructs. We have to come up with various ways of measuring them, of defining them, but we can't specifically observe them. Let's say I'm interested in love. How can I observe love? I can't. I see a boy and a girl rolling around in the grass outside. Is that love? Is it lust? Is it rape? I can't tell. But I define love by various specific behaviors. Nobody thinks any one of those in isolation is love, so we have to use a number of them together. Love is not eye contact over dinner. It's not holding hands. Those are just manifestations of love. And intelligence is the same."

The solution to the problem of measuring something that can never be directly observed, he explained, is to take multiple indirect measurements and then statistically calculate the degree to which they vary in sync with each other. Known in statistics as "latent variable analysis," the approach enables psychologists, economists, artificial intelligence researchers, and others to bring mathematical rigor to such otherwise fuzzy concepts as extraversion or introversion, quality of life, wisdom, happiness, and intelligence.

"What's really important is the variance," Engle explained. "Any one test doesn't tell you much. That's why, in my lab, we use at least three and sometimes as many as twenty different indicators for fluid intelligence, because we're looking for the factor that's common to them all, what these tests have in common when we remove the variance."

Although Jaeggi and Buschkuehl's first studies used only one or two measures of fluid intelligence, their latest had grown to include many more, at least in part to satisfy Engle's concerns. This explains why, when I finally sat down to have them measure mine, it took so painfully long.

•—•

Chris Cargill, an undergraduate working as a research assistant in Jaeggi and Buschkuehl's office, escorted me to a row of three tiny rooms. We entered the last one—the "green room"—just large enough for a molded plastic chair positioned in front of a computer resting on a countertop. I sat down in the chair, and Chris had to stand in the doorway as he explained that there would be six tests in all.

The first, called the "surface development," presents a series of what look like flattened pieces of cardboard cut in such a way that they can be assembled into oddly shaped boxes. The challenge is to figure out which sides of the flattened two-dimensional versions correspond to the assembled, three-dimensional versions. Chris stood by while I read the instructions and tried a sample problem. Then he wished me luck, told me I had six minutes, and closed the door.

Had I been escorted out to the pitcher's mound in Yankee Stadium and asked to pitch to Derek Jeter, I couldn't have felt more incompetent. After six minutes of struggle, the door opened.

"Whoa!" I moaned, forcing a few chuckles to sound like I was just being ironic. "That's the worst!"

"That's not an easy one," said Chris, in the seen-it-all-before-but-trying-to-be-nice tone of a cancer surgeon. "But none of them are. So this next one is called the APM. You've got this matrix puzzle where the bottom right piece is missing. Your job is to figure out the pattern so you can figure out which is the missing one."

"Oh, is this the Raven's?" I asked.

"Right, the Raven's advanced progressive matrices—the APM. So there's some examples here. Which do you think is the missing one?"

We went through the examples, he told me I could take as long as I wanted, and then the battle began. Actually, the first few problems seemed easy, even obvious. The next seven or so were difficult but doable. And then I hit the wall. I actually burst out laughing at one; the mysterious grid of symbols just looked like a spoof, as if it had been concocted by the staff of *The Onion*. And then it occurred to me: this must be what my dog feels like when she watches us talking, seeing our mouths move and sensing that something meaningful is coming out of them, but unable to figure out what.

Nearly an hour of struggle later, I walked out to tell Chris I was done. He then instructed me in how to take the next test, the digit-symbol; then another where I had to figure out which little shapes could be assembled into a bigger shape (the "space relations"); and then some other variety of mental torture (the "form board") that is probably outlawed by the Geneva Conventions, or should be.

"You need a break?" Chris asked, returning to the room.

"I need a drink."

With one last test to be administered, we decided to leave it for the afternoon. It was after 1:00 p.m. by the time I returned to the basement for my final assessment of fluid intelligence. This last one would be another matrices test—similar to the Raven's, but more difficult—called the Bochumer Matrizen-Test, or BOMAT. With Chris having left for the day, Buschkuehl began explaining how it worked. I was nearly finished with the sample problems and was just about to begin the formal test when the office went pitch-black.

"Not again!" said Buschkuehl. "We had a blackout like this a couple months ago."

Jaeggi and their graduate students wandered out from their offices. "It will probably last only a few minutes," she said. "Last time this happened the lights were back on in ten minutes."

Without any windows, the only light came from a few laptop screens. We stood around chatting, waiting for the juice to come back on. After about fifteen minutes, a man in overalls came down the hallway to announce that everyone had to leave the building.

"Did you have to leave the building last time?" I asked.

"No," said Jaeggi. "But it will probably be fixed in a few minutes."

We walked upstairs to the main floor and onto the street. Standing around, we discussed how absurd it would be if, after flying from my home in New Jersey all the way to Michigan, I couldn't finish the battery of tests I needed to take in order to begin my training regimen.

But that was exactly what happened. With my flight out of Detroit departing at 6:00 p.m., I needed to leave Ann Arbor by 3:00. We walked over to a café and chatted for an hour, and then it was time for me to leave.

Jaeggi and Buschkuehl apologized, but of course it wasn't their fault. And luckily, they were planning to move from the University of Michigan at the end of the year to take teaching positions at the

University of Maryland—just a five-hour drive from my home. So my plan to get my fluid intelligence measured before beginning a training regimen hadn't been ruined by the blackout, only delayed.

•—•

I finally did take that final test of fluid intelligence a couple months later, down in Maryland. Buschkuehl told me I could take as long as I wanted to on the BOMAT. I ended up spending about an hour and a half, much longer than I had spent on any of the other fluid-intelligence tests back in Michigan.

But my testing was not complete. Even though standard, old-fashioned IQ tests also include measures of crystallized intelligence, like knowledge of synonyms, I wanted to take one anyway, just to be thorough. So I turned to Mensa, "the high IQ society." Once a year, the local branches of Mensa in communities across the United States administer an IQ test to anyone interested in joining the society and willing to pay the $40 fee. Those whose IQ is found to be in the top 2 percent of the adult population are qualified to join.

But there was a catch.

Mensa does not permit applicants to take its IQ test twice, because the organization ascribes to the view that intelligence doesn't change much, so there's really no point in retaking it. Either you got it or you don't. And they require test takers to show a license or other official photo identification. That put a crimp in my plan to take the same test before and after my training regimen, until I searched my soul for moral scruples and decided they were more general guidelines than rules. Two key words came to me: Identity. Theft.

Back in the fall, I had flown out to Wisconsin and taken a ferry to Washington Island, off the northern tip of Door County, to hang for a long weekend with some of my friends from Beloit College. And during the visit, it seems that some beer was imbibed, and then some whiskey, and then the next thing I knew I was sitting in a lawn chair on the shore of Lake Michigan, smoking a cigar, while my dear pal Walt Roberts shaved my head neat as a bowling ball.

Stay with me, now. When I returned home, I realized that a friend whom I will call Richard (whose true identity I am sworn to withhold, lest I open him to criminal prosecution by the crack Mensa legal team) also had a beautifully shaved head, and also wore glasses, as do I. And so I sent him an e-mail, asking if he would allow me to borrow his driver's license when I took the first Mensa test.

"That is certainly one of the strangest requests I think I've ever received," Richard replied, "and it does beg the question, 'Can't your teenage daughter get you a fake ID?' "

But he agreed to do it for the sake of science. And so it was that on a frigid night in 2011, as I drove to take the big test, I was listening to my car radio when a report came on that a group of high school students on Long Island had paid people to take the SAT for them, even going so far as to use fake IDs. An officer of the College Board, which administers the test, said he didn't think that impersonation is a major problem for standardized tests.

As I drove, I wondered whether, if I got smarter through my training regimen, would that mean I would actually be able to write a better book? Would I be better at playing, say, chess? There was something unnerving about this idea of becoming smarter. Was it really possible? It all seemed insane, like I was preparing to jump out of a plane without a parachute.

At the building where the test was administered, in an ordinary corporate meeting room, three other people were taking it that night: two men, one woman, all of them looking decidedly nongeeky. Whatever stereotype I had imagined of Mensa members, none of these people fit it.

"Okay, Richard, are you ready to go?" The proctor was looking at me. I was unprepared for how disturbing it felt to be addressed with somebody else's name.

"Ready," I said.

After giving careful instructions, he told us to begin the first of seven subtests. The math challenges left me feeling almost unprepared and ignorant, but for the vocabulary test, on the other hand, my years as a journalist gave me what seemed like an unfair advantage. Words are literally my business. Asking me vocabulary questions is like asking a car mechanic about tools and auto parts.

Altogether, the tests took less than ninety minutes. Filing our test sheets into his briefcase, the proctor said that when he had taken the test years earlier, he had done terribly at the vocabulary questions, but really well on the math. That was the point of there being a variety of tests, he said, so that people with different strengths, math or verbal, could still compare their overall intelligence. Then he gave us all a Mensa pencil with the group's logo on it.

"Good luck, Richard," he said to me.

Driving home, I wondered whether my training regimen might work well enough that I would eventually be able to qualify for Mensa membership.

One last test before my training regimen could begin: an fMRI brain scan. Many studies over the years had established that size does matter when it comes to intelligence, but just a bit. In fact, about 6.7 percent of a person's fluid intelligence can be explained by the amount of "gray matter," the overall volume of neurons, in the brain. An additional 5 percent of fluid intelligence can be explained by the size of a particular region called the left lateral prefrontal cortex—a section located behind the upper-left edge of your hairline, and which becomes highly active during tests of working memory.

These modest but significant effects of particular brain regions offer insight into why women, whose total brain size volume averages about 10 percent smaller than men's, are nevertheless just as smart as men on average. Women actually tend to have more gray matter than men, whereas men tend to have more white matter than women. Men have generally been found to perform better on visuospatial tasks, while women generally beat men in verbal fluency and long-term memory. Although a 1983 study found that the proportion of boys scoring above 700 on the math portion of the SAT was thirteen times higher than that of girls, that ratio plummeted to less than four times higher by 2010, evidence of how societal attitudes and educational opportunities are at least as powerful as biological destiny. It remains the case, however, that males are slightly more likely than females to be cognitively disabled.

More important than size, researchers are finding, is the function of brain regions and how they communicate with each other. In August 2012, a study by researchers at Washington University in St. Louis examined the strength of the connections between the left lateral prefrontal cortex and the rest of the brain. While the *size* of the left lateral prefrontal cortex accounts for about 5 percent of fluid intelligence, the strength of its connections to the rest of the brain, they found, accounts for 10 percent, higher than any other observable brain factor.

"We had this hypothesis that if activity in this area is so important to intelligence, it's because it has to connect to other parts of the brain, to your perceptions and your memories and everything else," said Michael Cole, the first author of the study, a postdoctoral researcher in the Cognitive Control and Psychopathology Laboratory at Washington University, when I reached him by telephone. "We tried to see if we could predict or correlate individuals' fluid intelligence with how strongly their left lateral prefrontal cortex communicates with the rest of the brain while at rest, while the person isn't doing any particular task in the fMRI. Lo and behold, we could. It was highly statistically significant."

I asked if he would be willing to do a similar scan of my brain before and after I completed the brain-training regimen that I was planning. He said he would, but that I would have to check with his boss, Todd Braver, codirector of the lab he worked in. Even then, he added, they would have to get approval from the Ethics Review Panel at Barnes-Jewish Hospital, where their fMRI is located.

By the time everything was set, nearly a year had passed since I'd visited Jaeggi and Buschkuehl in Ann Arbor. Finally, on October 3, 2012, I arrived in St. Louis for my scan. Todd Braver greeted me outside his office and suggested we walk to a nearby café to discuss his and Cole's research. Only a little older than Jaeggi and Buschkuehl, Braver was already codirector of a prominent academic research laboratory. I began the interview by asking who his fellow codirector was.

"Deanna Barch," he said. "Actually, she's my wife."

Braver offered a totally objective description of her. "My wife is Superwoman," he said. "She is the most amazing researcher. The only downside of our relationship is that it's easy for me to feel insecure in comparison to her. She's probably going to be chair of our department in a couple years. She's the editor of the journal *Cognitive, Affective, and Behavioral Neuroscience*. She's involved with basically every committee possible at the university. She's the kind of person who wakes at four in the morning and writes a couple manuscripts before the kids wake up at seven. And she's the leader of our daughter's Girl Scout troop. She's just the kind of person who makes everybody feel like: What am I doing wrong? And she's an incredibly nice person, too. She's not arrogant or egotistical. And she picked me. That's the one thing that stops me from feeling insecure. She must have seen something."

He kept going on about her, but let's circle back to how it relates to the nature of intelligence. And yes, it actually does.

"I'm really interested in the connection between motivation and cognition," he said. "I got interested in understanding my own level of motivation; I worry sometimes that I don't feel motivated enough. I was the kind of kid, when I was six years old, people would say, 'This kid is a genius.' Both of my parents were professors. I come from the classic eastern European Jewish tradition. Everybody in my family has an advanced degree. There was no question I was going to go to college and get an advanced degree. And then I had a bit of a crisis and dropped out of college for a couple years. I had been doing everything just because it was expected of me."

By the time he'd discovered his own passion for psychological research, he was a graduate student in the laboratory of Jonathan Cohen, at Princeton, one of the world's leading researchers in the neurological underpinnings of cognitive control—the same subject that Jonides has devoted his career to.

Because every psychologist I have spoken to puts a slightly different spin on the definition of cognitive control, I asked Braver for his version. "When we're talking about cognitive control," he said, "it means being able to inhibit distractions, to maintain information in working memory, to switch between tasks and selectively attend in the face of interference. But we are talking about goal setting as well."

It is here, at the level of cognitive control, Braver said, that his wife and codirector really shines.

"What Deanna really has is cognitive control more than raw intelligence per se," he said. "Nobody tagged her as brilliant from early on. But this is where cognitive control and intelligence are related but not identical. She just has a very good sense of self-identity, a very good sense of delayed gratification, that immediate rewards tend to be less valuable than later rewards, and understanding what it takes to regulate impulses—impulse control is a big part of it. Deanna can concentrate amazingly well, and she can multitask when she needs to, and she also has really good cognitive control over her emotions. All of it creates a virtuous cycle."

Control of one's thoughts, control of one's emotions, control of one's goals and behaviors: Braver left me with the understanding that they are closely related, with underlying neural mechanisms that often overlap. And he also left me with the understanding that he is one of the most happily married men in North America.

Then it was time for my brain scan.

We drove to Barnes-Jewish and took the elevator to the tenth floor, home to the Center for Clinical Imaging Research. There we met Mike Cole, a couple of research assistants, and the operator of the fMRI, a big, silver-haired, goateed guy wearing the kind of long white jacket that scientists wear in laboratory settings. The fMRI control room, where they sat, had eight computer screens, some attached to laptops, some to older desktop units, and a glass window looking onto the adjoining room in which the fMRI was located. They instructed me to remove anything metallic from my pockets, as well as my glasses. Because I needed the glasses to see at a distance, and because the test would involve my viewing a computer screen while doing N-back exercises, they gave me a set of vision-corrected plastic goggles to wear and then brought me into the fMRI room.

About ten feet wide and high, the tunnel-like machine had a plastic panel sticking out of its mouth, on which I was told to lie. I felt no worries about getting slid inside the machine like a tray of cookies into an Easy-Bake Oven; claustrophobia was one fear I'd never experienced. So I lay back on the panel with the goggles over my eyes and a set of fat earphones around my ears (in order to hear their instructions from the other room), eagerly looking forward to the experiment. Then one of them said, without warning, "Okay, we're going to lock your head in now," and proceeded to lower a bridgelike device over my forehead and click it into place on the platform beneath me. The device had some kind of little mirror on it that went in front of my eyes, blocking my vision of the ceiling. It was meant to enable me to see directly behind me, where a computer screen was positioned at the back end of the machine, so that I could see the N-back task. But with my head immobilized, and the goggles and earphones surrounding my eyes and ears, I felt disoriented and panicked. And this was before they had even slid me inside the machine.

"Um, could you let me up for a minute?" I asked.

The assistant unlocked my head. I sat up, removed the goggles and earphones, and took some deep breaths.

"I wasn't prepared for my head being held down like that," I said. "That is intense."

I waited a minute, allowing the feelings of panic to give way, and then tried it again. This time I lasted about a minute before asking them to let me up. Waited another five minutes to calm down, then tried again, going long enough to let them actually have the machine begin sliding me inside the scanner. Made it halfway before I freaked again and had to be let out.

"I feel like such a failure," I said. "I can't believe I came all the way to St. Louis to do this."

The big guy who operated the device was super nice. "Don't worry," he said. "Lots of people get frightened. Usually they can take a sedative, but not during a working-memory test. Do you want to take a walk? Get a drink of ice water? Come on, I'll take you."

We padded down the hallway to a little kitchen area, where he got some ice and poured me some water. I drank it. "It's not being inside the machine that's really frightening me," I said. "It's having my head locked down with those goggles over my eyes and those big earphones. It's so disorienting."

"It happens every week that somebody can't handle it," he said. But I knew that none of those people were flying halfway across the country to do it for a book. We walked back to the room with the fMRI and I decided to walk around the machine, to orient myself as fully as possible to what would be happening. I went to the back, where the computer screen was positioned. I put the goggles back on, then the earphones.

"Okay, let's try this again," I said. "But this time I want to just lie on the platform for a minute or two with my head locked down and try to get used to it."

Once they had locked my head down, I brought my hands up to touch the goggles, the earphones, and the locking device over my head. I moved my feet around a bit, rotated my ankles, and tried wiggling my head up and down, discovering that it was actually able to move a fraction of an inch—not that I would want to do so during the scan. "You're perfectly safe," I told myself. "You're choosing to do this." When the urge to sit up welled up within me, I imagined myself lying on my bed with our dog, a crazy but sweet little Bichon named Sugar, lying in my lap.

"What do you think?" asked the device operator, standing nearby.

"Let's do this," I said.

He left the room and turned the switch to send the platform sliding into the doughnut hole.

"You okay?" he asked, his voice coming through the earphones. "Want me to start 'er up?"

"Start 'er up," I said.

A terrific throbbing noise, with a staccato beat like a jackhammer or speed-metal drumming, enveloped me for the next three minutes. After thirty seconds of silence, the next sequence sounded eerie, with a bzzzz-pause, bzzzz-pause, bzzzz-pause, evoking the soundtrack of a 1950s sci-fi movie when a ray gun is vaporizing a station wagon. After another moment of silence, a different kind of noise erupted, this one more of a rhythmic thumping, like an out-of-control washing machine, but with a hollow "pock" sound coming at the end of each thump. Finally, after a ten-minute session of wailing and grinding and gnashing, they were ready to have me do an N-back exercise as they continued scanning, the noise of the machine greatly reduced. Unlike the one that Jaeggi and Buschkuehl used, this N-back version presented a series of images, and the task was to press a "target" button they had placed in my hand if the exact same image repeated from two times before. Sometimes the images were faces, sometimes tools, sometimes photographs of building facades, sometimes landscapes, and sometimes odd close-ups of arms, hands, and legs. Some of them were obviously meant to look misleadingly similar to each other.

At first I was totally confused—I had thought that all N-back tasks were the version that Jaeggi and Buschkuehl used, with black dots moving around a tic-tac-toe grid. But after getting over the initial surprise, and realizing that I had to pay very close attention to the faces or tools or whatever in order to keep track of them, I thought I got the hang of it pretty well.

And then, after about forty minutes inside the machine, it was over. Back in the control room, they showed me some images of my brain.

"So I do have one," I said. "That's reassuring."

My pre-testing complete, I was now ready to begin my brain-training regimen. But how exactly does one go about doing that? *What really works?* Does anything other than the N-back have truly solid, scientific evidence behind it for making people smarter?

CHAPTER 3

A Good Brain Trainer Is Hard to Find

There turns out to be a large number of commercially available programs claiming to have scientific evidence supporting the effectiveness of their brain training. But figuring out which ones stand up to scrutiny, and which if any I should include in my own regimen, was not easy. Perhaps the best-known, first released in 2005, is Dr. Kawashima's Brain Training, also called Brain Age: Train Your Brain in Minutes a Day! Although over 19 million copies of the software have sold, and even some neurologists have been known to recommend it for prevention of Alzheimer's, the maker of the game, Nintendo, insists the software is purely for entertainment and declines to support any benefits. Only a few studies, in fact, have ever found such benefits.

But five commercially available approaches do make plausible claims of effectiveness, including one developed by Torkel Klingberg, the Swedish scientist whose 2002 study inspired (and baffled) Jaeggi and Buschkuehl. By the time his study was published, Klingberg and some colleagues had already partnered with the Karolinska Institute to form a company, Cogmed, to turn working-memory training into a business. They brought a measure of scientific credibility to the enterprise that had previously been lacking. Without the hyperbole of self-help hucksters, Cogmed insisted on offering its computerized training through psychologists and other clinically trained PhDs. The initial target market was children with ADHD, whose parents hoped to find something other than drugs to improve their children's attention.

By 2003 Cogmed already had its first paying customers in Sweden. Two years later, a second, larger study by Klingberg and colleagues found that their working-memory training improved complex reasoning and parental measures of behavior in children with ADHD. In 2006 they recruited and trained four psychologists in the United States to offer Cogmed, as well as one in Switzerland. By 2010, psychologists in twenty-five countries on six continents offered the training, and additional studies had been published showing it to help both children and adults with a variety of cognitive disorders. That same year, in a major step suggesting just how vast a business this brain training could be, Cogmed was sold to Pearson, the largest education company in the world. (Full disclosure: Pearson is also co-owner of the largest book publisher in the world, Penguin Random House, which in turn owns Hudson Street Press, this book's publisher in the United States.)

"We work both with healthcare providers and schools all over the world to provide Cogmed training to the many people who are constrained by their working memory," the company states on its website. "Cogmed is a leader in the emerging field of evidence-based cognitive training. We have scientifically validated research showing that Cogmed training provides substantial and lasting improvements in attention for people with poor working memory—in all age groups. That makes Cogmed's products the best-validated products on the market."

Aren't claims like this rather overheated? I posed that question to Klingberg at a tiny, crowded coffee shop on West 23rd Street in Manhattan, where he was in town to give a talk at Columbia University later that afternoon. He wore a black leather jacket and a momentary scowl, having heard such critical questions many times before.

"Yeah, well," he said with a shrug. "We did start to do research in 1999. Of course you can say that we still don't have—we should wait another ten years until we have thousands of participants. This is a general problem with cognitive training studies that we don't have these huge studies that drug companies have. On the other hand, this is not something that is dangerous. If we have worked for five years, should we allow people to try this? I think so."

He pointed out that Cogmed claims only that it can improve working memory, not fluid intelligence per se—even though many studies have found that working memory and fluid intelligence are closely related.

"What we see, over and over again," he said, "is improvement of working memory and also of attention, including attention in everyday life. This is not everything, but it's good enough for me if we can have that. Working-memory problems and attention problems are huge for many children and adults. Right now I don't have any financial interest in Cogmed anymore. The influence I have had over Cogmed has been to make them very cautious. They don't make claims about rejuvenating your brain or improving intelligence."

Although I had read all his studies, I told him that I was still having a hard time understanding exactly what kind of computerized training tasks Cogmed offers.

"There are twelve tasks," he said. "They're all visuospatial. The role of attention in working memory almost always has a spatial dimension. When you're paying attention—even when you're paying attention to me talking here in this café—there's a spatial component.

When there's a loud noise, you might shift your attention to where it's coming from. Being able to maintain your spatial focus on me is important for you right now. Even though it's words coming from me, there's an important component of space. So if you can improve the stability of that spatial aspect, you will be better at visuospatial tasks and be better at keeping your focus on me rather than on that noise over there."

Still wanting a better sense of the exercises Cogmed offers, I scheduled a meeting with a clinical psychologist, Nicole Garcia, who offers the training just a few miles from my home in Montclair, New Jersey. She allowed me to sit down at her computer to play a handful of the games. (She emphasized that Cogmed calls them "training tasks," not games. But they looked like games to me.)

I clicked one called 3D Grid. It showed the inside of a cube, looking down on it as though the top were removed, with each of the four sides and the bottom divided into four panels. Once the game began, some of the panels lit up; I then had to click on them in sequence to show that I remembered. Another game, called Hidden, showed a standard numeric keypad, the kind used on cell phones and calculators. The keypad was hidden while a man's voice recited a short list of numbers. When his list was complete, the keypad reappeared, and I was supposed to click on the list—in reverse order. A third game showed a circle with nine smaller circles strung along it like pearls on a string or carriages on a Ferris wheel. As the big circle slowly rotated in a clockwise direction, the little circles lit up in a random sequence. Once the sequence was completed, I had to click on the little circles in the same order.

All of the games were ridiculously easy on the first pass, but immediately grew hard enough—meaning that the sequence to be remembered grew longer and was presented faster—that I began making mistakes.

"Doing these tasks yourself is the only way you can understand how difficult it can get—and it gets very, very hard," Garcia said, laughing. "I had to do all twenty-five training sessions, so I know."

Although she had heard about Cogmed back in 2004, she said she waited to get trained and didn't start to include it among the treatments she offers children and adults until 2011.

"I had a client who was twenty-three and had been dealing with ADHD his whole life," she told me. "He was on ADHD medication, and it was helping, but we reached a point where I felt like we were hitting a wall. He was still in college and really having trouble completing his classes. I looked at different kinds of programs and ended up deciding to pursue Cogmed. I thought this might be the thing that opens the door for him so he can move forward."

Over a year since he completed his training, she said, he continues to see benefits. "Cogmed got to a part of him that I could not get to as a psychologist with talking, and that the medication didn't affect either. This is the first semester that he hasn't dropped out of any classes and he hasn't gotten any Fs."

Having offered Cogmed to a few dozen of her patients, as young as six and as old as sixty-three (including a successful attorney whose ADHD was diagnosed in her forties), she said she's convinced of its benefits, sometimes in combination with medication and sometimes on its own. Although the training is all completed on a computer, she emphasized that Cogmed requires that a trainer work with each person, to maintain motivation and perseverance. And at a total cost to families of about $2,000 for the twenty-five sessions, she said, it compares favorably with many other kinds of ADHD treatments.

"Not everyone is appropriate for it," she said. "But when they're able to actually do the exercises and complete the training, I have not seen anyone who hasn't benefited. That includes me. I have always had the worst sense of direction. Everyone has deficits. Even with my GPS, before Cogmed, when it would say 'at the next street, turn right,' I would miss the turn. I had no idea my driving would improve because of Cogmed, but one day I was—Oh, my God, I'm not getting lost anymore."

While testimonials like Garcia's are colorful, to skeptical scientists they are the least convincing kind of evidence, particularly since every screwball form of treatment ever offered, going back thousands of years throughout the history of medicine, has relied on testimonials. "See, the doctor gave me the leeches and now my carbuncles are cured!" (Which is kind of a funny example, because leeches have actually made an unexpected comeback as a sound treatment for cleaning certain kinds of wounds.) Where Cogmed beats all other forms of cognitive training is in the number of published, randomized clinical trials demonstrating its benefits and the number of trials still under way, led by independent researchers at leading institutions without any commercial connection to the company.

"I came in as a skeptic," said Julie Schweitzer, director of the ADHD Program at the University of California in Davis's MIND Institute. "My concern was that you have parents who are desperate, who are not satisfied with the treatments their children are getting, or maybe they're not comfortable with medications."

When she wrote a grant application to study Cogmed as a treatment for ADHD, she told me, "I was shocked that one of the reviewers' critiques was, 'We already know that this works.' I felt, and still feel, that a lot more research needs to be done."

Her study of twenty-six children diagnosed with ADHD compared twenty-five sessions of Cogmed training to another kind of computerized training that does not get progressively more difficult as children's abilities improve. To test its effect on one of the most serious consequences of ADHD—the inability to remain on-task in the face of distractions—she used an objective measure called the restricted academic situations task, or RAST.

"We put the child in a room, give them some toys to play with, put the toys to the side, and then we give them a stack of math problems," she explained. "We tell them to work on the math problems for fifteen minutes and then we leave the room, but we're videotaping them the whole time. Later we score the videotape every thirty seconds: either they're on-task or they're off-task, whether playing with toys, vocalizing, fidgeting, getting out of their seat, or whatever."

When published in July 2012 in the journal *Neurotherapeutics*, Schweitzer's study found that children in the placebo group spent just as much time off-task at the end of the study as they had at the beginning, but those who trained on Cogmed sharply increased the amount of time they spent doing the math problems—over six minutes more on-task than the placebo group.

"We got positive results, but it was a very small study," said Schweitzer. "In general, I'm cautiously optimistic about the potential of working-memory training for ADHD. There's a huge need, especially with the adolescents I work with, for something more than the traditional treatments. Medication is the best thing we have for now and it's extremely effective, but it doesn't fix everything. A lot of my patients are either no longer willing to take it, or it may have too many side effects. We need more tools. If this works, it's fantastic."

ADHD is hardly the only disorder in which working-memory skills are adversely affected. Schweitzer is now running a small study of Cogmed for children with fragile X syndrome, a genetic form of intellectual disability. Although the study is not yet completed, she said, "I can tell you that there certainly are a number of kids who are able to do the tasks. The majority of them have been able to do it. The parents are very excited."

Children who have survived cancer are another group often in need of cognitive rehabilitation. "Somewhere around 20 to 40 percent of children treated for leukemia will end up with cognitive changes over time," said Kristina K. Hardy, a neuropsychologist at Children's National Medical Center in Washington, D.C. "For those treated for brain tumors, the figure is conservatively around 60 to 80 percent."

What distinguishes these young survivors from most others seeking cognitive rehabilitation is that the effects of radiation or chemotherapy on the brain become apparent only with the passage of time. Immediately following treatment, a recent study found, survivors of acute lymphoblastic leukemia showed no significant change in their verbal IQ scores, but by early adulthood, their scores had dropped by an average of 10.3 points.

"People with, say, traumatic brain injury lose a bunch of skills overnight, and you're hoping to build them back up," Hardy told me. "Our kids treated for cancer are not losing skills. They're just failing to gain new ones as well as they used to. When they first come in, we don't see many effects of the treatment. But then about a year later, when they're all done with treatment and back to school, they're not learning new material like they used to, they're not getting the same grades, because their working memory and their attention have been affected. Over time, you often see declines in academic functioning and in IQ."

Although many children do fully recover their pretreatment cognitive abilities without further treatment, for those who don't, the effects can sometimes be seen most fully in adulthood, Hardy said.

"When we look at adult survivors of childhood cancer who were treated long ago," she said, "we find as a group that they aren't achieving the developmental milestones at the same rate as their peers. They aren't getting married, they aren't moving out of their parents' houses, they aren't finishing school, they aren't getting jobs as soon or as often. As a clinician, I find it really depressing when you have a kid who was so unfortunate as to have a cancer diagnosis, and then survived that, but now has cognitive changes that affect his or her life forever.

And knowing what can happen, what's really sad is when I see a child who just completed treatment, and I don't have anything to do to prevent those changes. So I started to look at ways to improve cognitive functioning and improve those outcomes."

In 2012, she reported the results of a pilot study comparing Cogmed to a placebo form of computerized training that does not get increasingly difficult as the children get better at the tasks. Among twenty children who had survived either brain cancer or leukemia, those who trained with Cogmed saw substantial improvements compared to the placebo group on their visual working memory and in parent-rated learning problems. "I entered this work with some skepticism that just doing some computer work could help anybody," Hardy said, echoing Schweitzer's words. "I thought we wouldn't be able to move the needle. And not everybody has been able to make a lot of gains. But I've had some kids who not only reported that they saw big changes in their life after training, but when we brought them back to the lab and did neuropsychological testing, we saw great changes, too. At a gut level, I do believe some kids can do better with this training. I don't believe that every kid is going to. But in my most recent trial, we saw about 50 to 60 percent of children make gains that I consider clinically meaningful."

On the basis of the evidence she is aware of, I asked Hardy if she would generally recommend Cogmed to children who have survived cancer or are facing other cognitive challenges.

"I am extremely cautious about making any recommendations right now," she told me. "The numbers in my published study are small. We still don't know what the optimal amount of training is, or who might be most likely to benefit. But I'm continuing to study more children, and researchers at St. Jude's Hospital in Memphis, and at the University of Minnesota, are also running studies of Cogmed. I'm really excited to see the data."

In the meanwhile, she said, "For these kids, we're really at a place where we either sit back and watch them get worse or we try to do something. To me and to a lot of families, working on computer exercises for thirty minutes a day is something that feels very benign. At worst, we won't have helped them at all. But we know that changes in working memory are very closely linked to intellectual functioning and academic achievement over time. So it makes sense that if we can catch the working-memory problem as it's developing, we might be able to mitigate or delay these longer-term problems."

My bottom line: I was impressed by Cogmed's research, the seriousness of the researchers involved with it, the fact that it's offered by trained psychologists, and its fairly reasonable price. If I were looking to remediate a diagnosed cognitive disorder, I would certainly consider it as among my first choices. But since I'm looking to increase my fluid intelligence, it did not look like a good fit. For my own personal training regimen, I scratched Cogmed off my list.

Lumosity

A less medicalized, more democratic approach to cognitive training is taken by Lumosity, whose television commercials have helped propel it to 40 million members, although the company does not disclose how many are paying subscribers. Two neighbors on my block told me they use it, and a week before I flew to San Francisco to visit the company's headquarters, I received an e-mail from another old pal from Beloit College. "My sister is trying to get me to sign up for this 'Brain Training' system," he wrote. "Have you heard of these guys?" The link was to Lumosity.

Arriving at their sixth-floor office in a rehabbed building on Kearny Street in San Francisco's downtown, I walked past an exposed-brick wall into a large open space where a couple dozen twenty-somethings tapped intently at their computers without even cubicle dividers between them. A kitchen area had three coffeemakers on the counter and two glass-doored refrigerators—one stocked with juices, the other with beer, wine, and vodka. It fit my image of what a satellite office of Google would look like. Except, oh, wait a minute, that crossword puzzle taped to the wall? It was made by Tyler Hinman, who actually quit Google in 2010 to work here, after winning the American Crossword Puzzle Tournament five times in a row.

"I looked around for a place that would get me closer to the kinds of games and puzzles I enjoy," Hinman told me. "But where crosswords and Sudoku are intended to be a diversion, the games here give that same kind of reward, only they're designed to improve your brain, your memory, your problem-solving skills."

The company's press representative ushered me toward what looked like a metal garage door, wide enough for two cars; a button was pressed and it rolled up into the ceiling, revealing a conference room. We were joined by Michael Scanlon, the company's cofounder, and Joe Hardy, vice president of research and development.

As recently as early 2005, Scanlon told me, he had been pursuing a doctorate in neuroscience at Stanford University. "I was looking at African cichlids—they're really cool little fish," he said. "If you take one male and put it into a tank with another male, suddenly one of them will no longer be able to reproduce. Its gonads shrink; it goes from bright colors to gray. It's really neat, because you can watch it happen in minutes and hours. I was looking at what brain changes take place to produce those changes in the reproductive system. It was really interesting, but after a while I started to feel like it was pretty distant from anything to do with people."

A friend of Scanlon's from Princeton, Kunal Sarkar, had worked for a private equity firm that reaped profits by investing in 24 Hour

Fitness, a gym chain. Sarkar was looking for another business the firm could invest in, something like 24 Hour Fitness, but for the mind. He couldn't find anything, particularly not with the kind of inviting, inclusive, mass-oriented vibe that distinguished 24 Hour Fitness. So he, Scanlon, and Dave Drescher, a programmer and game designer, decided to build one themselves.

To devote the necessary time to getting it off the ground, Scanlon left his cichlids at Stanford in 2005 with only enough savings to last about a month and a half.

"So I started playing online poker," he told me. "I was playing a couple of hours a night. I had a good friend who made over a million dollars in a couple of years. He was kind of my tutor."

The biggest gamble of Scanlon's life paid off when Lumosity launched in 2007 and began growing 20 to 25 percent every quarter. In June 2011, the firm received $32.5 million from a venture capital firm. Within a year, according to the company, their number of members had reached 25 million. By April 2013, they claimed 40 million members.

"Your brain—just brighter" is the company's slogan. When I first visited its website back in 2011, it stated: "Our users have reported profound benefits that include: Clearer and quicker thinking; Faster problem-solving skills; Increased alertness and awareness; Better concentration at work or while driving; Sharper memory for names, numbers, and directions."

Those results are achieved, the company's founders say, by dressing up boring but well-established cognitive tasks, including the N-back and the complex span, into colorful games. Their version of the N-back, in fact, is unrecognizable from Jaeggi and Buschkuehl's. (Although the two researchers did advise the company on the game's construction, they did so without any compensation whatsoever, they told me.) The Lumosity version shows a frog hopping among lily pads. Players begin by clicking on whichever pad the frog just left

(1-back), then graduate to having to click on the pad the frog was on two, three, four, or more hops before.

But Lumosity's roster of games includes many that have nothing to do with N-back. "For Monster Garden," said Hardy, "you see a bunch of monsters pop up on this grid. Then they go away, and you have to navigate a path through the grid without stepping on any of the squares where the monsters were."

Although some of the games look indistinguishable from those offered by Cogmed, that is where the similarities between the two companies end. No psychologist is required to assess, monitor, or coach people who use Lumosity; anyone can sign up online, for a fee currently priced at $14.95 per month or $79.95 per year. And not all of the forty-plus games, as best as I could tell, are based on published scientific evidence showing that practice will translate into useful real-life improvements.

That said, Lumosity has already been the subject of fifteen studies published in scientific journals or presented at scientific meetings, and continues to be used in dozens of ongoing studies. In 2011, Shelli Kesler, a psychologist and assistant professor at Stanford's Center for Interdisciplinary Brain Sciences Research, published two pilot studies of Lumosity. One involved twenty-three children who had survived cancer, finding that practice on Lumosity improved their mental speed, flexibility, and memory. Another involved sixteen girls between the ages of seven and fourteen with Turner's syndrome, a genetic disorder associated with cognitive deficits in math. Training on three math games that Kesler developed in collaboration with Lumosity was found to improve the girls' math skills. Neither study was randomized, so it was impossible to know if the changes were due to a placebo effect. But in May 2013, Kesler published her latest study of Lumosity, this time using a randomized design involving forty-one breast cancer survivors, twenty-one of whom were assigned to train for forty-eight sessions over twelve weeks, while the remaining twenty were put on a waiting list. Kesler reported in the journal *Clinical Breast Cancer*: "Cognitive training led to significant improvements in cognitive flexibility, verbal fluency and processing speed, with marginally significant downstream improvements in verbal memory as assessed via standardized measures. Self-ratings of executive functioning skills, including planning, organizing, and task monitoring, also were improved in the active group compared with the wait list group."

"One of the reasons I like Lumosity is the way they offer lots of different kinds of tasks rather than just one thing," Kesler told me. "There's pretty strong evidence that several different types of medical problems, including cancer, HIV, diabetes, and MS, are associated with significant cognitive issues. I see these patients in clinic, and I almost always recommend that they try Lumosity or something like it. Fundamentally I do believe it helps. Even if you're a healthy person, this is a good thing to consider doing, to keep your brain active and healthy. My only hesitation is that if you start claiming it will work for everyone in every situation, that's where you can get in trouble."

Although the company actively supports continued research to document precisely which games help which people in which kinds of situations, the CEO, Sarkar, told me he simply finds it unimaginable that people's cognitive abilities cannot be improved through training.

"I ran very seriously in high school," said Sarkar, who grew up in Nagpur, India, a small town deep in the country's interior, until his father, a civil engineer, moved the family to Long Island when Sarkar was twelve. "When I showed up for the first day of track practice, I could see which kids were pretty fast and which were not. But who we are is not predestined. I was not the fastest kid in my freshman year of high school, but I worked really, really hard. I probably never could have been the top collegiate runner because of my talent level, but I certainly got to a fairly good place. So I feel like this basic notion that you can't get better, that something is fixed and completely unchangeable—that's not true. If I believed that, I wouldn't be here. It's true we're in the early innings of learning how cognitive training works, and we're making substantial investments in building the evidence. But what we're excited about is this opportunity to create products that actually help."

Already the size of the company's database, charting the progress and demographic variables of each of its users, is the envy of many academic researchers.

"We have the world's largest database of human cognitive performance," Joe Hardy told me. "Google knows what you're likely to buy. But we know what makes you smart."

In March 2012, the firm released its first public analysis of that database, showing that users who reported getting seven hours of sleep or having one or two alcoholic drinks per day performed better on cognitive tasks than those who had either more or less sleep or alcohol. That's right: moderate drinkers performed better than teetotalers, and sleeping too much proved just as harmful to mental performance as sleeping too little. The greatest effect, however, was from exercise. Those who worked out at least once a week performed 9.8 percent faster, solved 5.8 percent more math problems, and had 2.7 percent better spatial memory than those who never exercise.

Another surprise from the company's database was the age range of its users. Although Scanlon and Sarkar conceived of Lumosity as appealing to aging baby boomers, one-quarter of its audience turned out to be made up of students between the ages of eleven and twenty-one.

"The particular audience I had in mind at the earliest stages of the company," Scanlon confessed, "was my mother."

Perhaps the biggest challenge facing the company is not providing more evidence that its games truly do improve cognitive functioning—

40 million members apparently already believe that to be true—but keeping those customers coming back and actually practicing. That explains why gamers like Tyler Hinman work alongside neuroscientists and psychologists.

"A ton of our effort goes into making it enjoyable, certainly more enjoyable than going to a gym," Scanlon said. "We have the same issue that gyms like 24 Hour Fitness have. People join with the best of intentions, are feeling good about their progress, and then life gets in the way. So a big part of our day-to-day work here is to make the product as engaging as possible, to improve the compliance rate."

My bottom line: Although not all of Lumosity's games can boast evidence supporting their value on par with the N-back or Cogmed's working-memory tasks, many of them can. And that same variety may well be an asset, particularly since the scientific community has yet to agree on exactly which tasks are most useful for any particular person. No study has yet compared the effectiveness of, say, N-back to the complex span tasks offered by Cogmed. So for all we know, the smorgasbord of forty-plus games might include some that will be found to be more powerful than anything offered by Cogmed. Although, frankly, some will surely be found to be dogs that don't fetch. Still, the price is right—I've spent more at Starbucks in a day than I have for my monthly Lumosity fee—and I also liked the fact that it's available twenty-four hours a day.

With some trepidation, I added Lumosity to the plan for my regimen, along with the N-back.

Posit Science

Two blocks from Lumosity's headquarters, in a suite of drab offices that look like they could be an accounting firm's, I found a far more sober and medically oriented approach taken by the pioneer of the field, Michael Merzenich, the neural plasticity researcher whose early studies inspired Torkel Klingberg. One of the most prominent neuroscientists of the late twentieth century, Merzenich played a key role in the development of cochlear implants for people with severe hearing impairments. Born in 1942, old enough to be Scanlon's grandfather, his hair turning from gray to white, Merzenich is the crotchety éminence grise of the field.

"If you're not satisfied by some of these studies showing that training can be effective, you're an idiot," he told me. "If you just look at the number of controlled trials that relate to whether you can train the brain, the evidence is massive. It's no longer controversial. It's only controversial to a know-nothing. But we're still a drug-bound society. There's this strong belief that everything is going to be done with a pill."

In 2007, Merzenich retired from his position as codirector of the University of California at San Francisco's Keck Center for Integrative Neuroscience to devote himself full-time to Posit Science, a company he cofounded in 2004. His focus, and therefore the company's, is applying training protocols for serious cognitive disorders, ranging from Alzheimer's disease and traumatic brain injury to even schizophrenia.

"We're initiating an FDA trial for treating schizophrenia, using the computer as a medical device," said Merzenich. "This is medicine. It is driving changes in the brain."

While antipsychotic medications have long been used to relieve the delusions and hallucinations that are the most striking symptoms of schizophrenia, it's the so-called negative symptoms of the disease—severe declines in working memory and simple reasoning skills—that can be most disabling and for which no drugs have been shown to be effective. This treatment gap is what Posit Science's training protocols seek to fill. Two recent studies led by Sophia Vinogradov, vice chair of psychiatry at UCSF, have shown that fifty to eighty hours of training significantly improved patients' verbal working memory and learning, their ability to distinguish reality from fantasy, and their overall social functioning up to six months later. Vinogradov is now leading three studies involving a total of 260 people, in hopes of meeting criteria set by the FDA for approval of the training program as a treatment for the disorder's cognitive symptoms.

"My view is that it has the strong potential to add a new treatment tool to what we are able to offer patients with psychiatric illness, using an approach that is radically different from what can be accomplished by medications or by psychotherapy," Vinogradov told me. "This form of intervention is directly targeting some of the key information-processing abnormalities that contribute to the illness, rather than just palliating symptoms, as medications do, or teaching more adaptive responses, as psychotherapy does."

On April 9, 2012, the National Institute of Mental Health convened a meeting in Bethesda, Maryland, to review the state of the science on using cognitive training to treat mental disorders. A report from the meeting concluded that more carefully designed research is needed, particularly on how the training improves people's everyday functioning. But it stated: "Recent reviews are encouraging in that they conclude that cognitive training programs result in significant, albeit modest, improvements in performance on neuropsychological measures of specific cognitive skills (e.g., memory, attention, problem-solving)."

Reviewing the actual tasks that Posit uses, I was struck by how similar some of them are to games offered by Lumosity. Where Posit challenges the user to follow the movement of two or three balls floating on the computer screen amid a bunch of other balls, Lumosity has orange fish swimming around an aquarium. And both offer a game where birds momentarily appear and disappear against a background, challenging you to see and click on one (often on the periphery of the screen) after it's gone; Posit calls its version Hawk Eye, while Lumosity calls its game Eagle Eye.

Where Posit goes into unique territory that other providers do not is with tasks that challenge a person's ability to quickly perceive visual or auditory gradients. With the "sound sweeps" task, you listen to tones that are rising or falling, like an ambulance siren approaching or receding. The faster the tones are played, the harder it is to distinguish whether they are rising or falling. At one hundred milliseconds, or one-tenth of a second, I found it easy; by thirty-four milliseconds I was struggling. My problem, according to Merzenich, is that my middle-aged brain perceives sounds (and sights) at a much slower rate than a younger person's brain. Like a digital photograph with too few pixels, or an MP3 with too few bits of digital sound, my brain is just not taking in enough information.

"Your brain's machinery is recording in a degraded way," he told me. "So you have to deal with the fundamental machinery. You have to refine these perceptual abilities by engaging them on the cutting edge of resolution. That's what these tasks provide."

I had a hard time understanding what any of this has to do with intelligence, until Merzenich explained that he sees the "sound sweeps" training as part of a continuum of tasks, from the simplest to the most complex, which all play off each other and all require fine-tuning.

"We're trying to assure generalization by first training the brain to make these elementary distinctions," he said. "We go from the most basic sound dimensions of phonemic distinction to vowels and consonants, to distinctions between words, to narratives, and ultimately to the manipulation of information in cognitive control. The goal is to drive these fundamental processes to correctly represent information with greater power and salience, and then trying to ensure that the person is using this in higher analysis and thought. The simple fact is in a normal adult life, you don't practice enough on the things that gave you the powers you gained during childhood. You become a user of mastered skills. You're not working to maintain the high abilities on which it was all based."

Evidence supporting his program's benefits for older people comes from three of the largest randomized clinical trials ever conducted of cognitive training programs. The Improvement in Memory with Plasticity-based Adaptive Cognitive Training (IMPACT) study, directed by researchers at the Mayo Clinic in Rochester, Minnesota, involved 487 people aged sixty-five or older, all living in their homes without any diagnosis of significant cognitive impairment. Eight weeks of training on Posit Science's programs one hour per day, five days per week, produced significantly better improvements in memory and attention than a placebo program did.

A second study, published in May 2013, involved 681 people in two age groups: fifty to sixty-four and sixty-five-plus. Compared to the active control group, assigned to do computerized crossword puzzles, participants of all ages who trained for ten hours with Posit's speed-of-processing program gained a wider "useful field of view," the kind necessary to see things in the periphery when driving, and also showed significant improvements on a number of untrained cognitive tests. Converted to years of protection against age-related cognitive declines—the "cognitive reserve" mentioned in this book's introduction—the results of just ten hours of training reflected up to six years of protection.

The third and largest study, the Advanced Cognitive Training for Independent and Vital Elderly (ACTIVE), involved 2,802 people aged sixty-five to ninety-four. It tested not only Posit Science's speed-of-processing training, but other programs designed to improve either long-term memory or reasoning. Its findings were mixed. Each training method improved people's abilities within the area trained, and Posit's speed-of-processing training group also showed statistically significant improvements on self-rated health and other measures that persisted for five years. But the training was by no means a panacea: during that same five-year period, a total of 189 participants developed dementia, and none of the cognitive training methods used in the study appeared to lower people's risk for it during that period. "Longer follow-up or enhanced training may be needed to fully explore the capacity of cognitive training in forestalling onset of dementia," the researchers concluded.

Before leaving Merzenich's office, I asked how he got into cognitive training from the seemingly unrelated pursuit of designing cochlear implants.

"The very fact that a cochlear implant works is the most beautiful example of adult brain plasticity that I can think of," he said. "You put a device into the inner ear that uses electrodes to represent audio information in a very different way than the brain normally receives it. There's no similarity between the way the brain is engaged by an intact ear and the way it's engaged by a cochlear implant. So what the brain hears initially sounds like crap, but eventually it sounds normal. The brain comes to understand it. That's the miracle of brain plasticity."

As I stood up to leave, I noticed something sticking to the leg of Merzenich's khaki pants. It was one of those adhesive length-and-waist-size tags that stores apply. Rather than be so rude as to point it out, I commented approvingly on his leather bomber jacket hanging from a hook on the door.

"I pay no attention to clothing," Merzenich told me.

My bottom line: Posit has scientific cred as strong as Cogmed's. But where Cogmed appears to be satisfied operating in the crowded marketplace for ADHD treatments, Merzenich is seeking FDA approval for treating the negative symptoms of schizophrenia—a daring step with historic dimensions if it pans out. Posit offers a version of its program online, at www.brainhq.com, in an attempt to reach Lumosity's mass market, but without its sophisticated design. Its studies of serious cognitive disorders, however, have all been carried out in the offices of psychologists, in a manner similar to Cogmed's. I urge anyone with a family member struggling with schizophrenia to look into Posit's training programs. But despite enormous admiration for all that Merzenich has done in the field, for me, for now, seeking to raise my fluid intelligence, I scratched Posit Science off my list.

LearningRx

The day after returning home from San Francisco, walking past a yoga studio that my wife, Alice, occasionally attended, I noticed a new sign over its front door, for a business called LearningRx. I walked in and picked up a brochure. "Train the brain, get smarter, guaranteed," stated the headline.

What the hell? This brain training was everywhere.

LearningRx turns out to be the most expensive, least supported by published research, and most aggressively marketed of the four leading cognitive training programs. It bears the curious distinction of being the only one set up as a franchise, like McDonald's, with independent owners running each of the eighty-three LearningRx centers in twenty countries. And neither the franchise owners nor the trainers who work for them are required to have anything more than a four-year college degree.

But hang on. LearningRx also has some unique assets, in particular that its training is offered in person, rather than on a computer, with a trainer encouraging each student to persevere—an important asset for children or adults struggling with issues of attention and focus. (Indeed, the problem of keeping people motivated to train on inherently challenging tasks is one cited by every researcher who has tested training methods, including Jaeggi and Buschkuehl.) Moreover, many of the tasks that LearningRx uses are the same kinds used by other cognitive trainers, except that they have been translated from a computer format to tabletop exercises performed with playing cards and other materials.

Early one Wednesday evening, with the permission of the franchise owner, I visited the LearningRx nearest my home. Nick Vecchiarello, a sixteen-year-old from Glen Ridge, New Jersey, sat at a desk across from Katie Duch, a recent college graduate who wore a black shirt emblazoned with the words "Brain Trainer." Spread out on the desk were a dozen playing cards showing symbols of varying colors, shapes, and sizes. Nick stared down, searching for three cards whose symbols matched.

"Do you see it?" Katie asked encouragingly.

"Oh, man," muttered Nick, his eyes shifting among the cards, looking for patterns.

Across the room, Nathan Veloric, twenty-three, gazed at a list of numbers, looking for any two in a row that added up to nine. With tight-lipped determination, he scrawled a circle around one pair as his trainer held a stopwatch to time him. But halfway through the fifty seconds allotted to complete the exercise, a ruckus came from the center of the room.

"Nathan's here!" shouted Vanessa Maia, another trainer. Approaching him with a teasing grin, she clapped her hands like an annoying little sister. "Distraction!" she shouted. "Distraction!"

There was purpose behind the silliness. Vanessa was challenging the trainees to stay focused on their tasks in the face of whatever distractions may be out there, whether Twitter feeds, the latest Tumblr posting, or old-fashioned classroom commotion.

Nathan has had learning issues since elementary school, his mother told me. He had recently graduated from William Paterson University with a degree in communications when she heard about LearningRx from a business networking group and thought it might help him in business and life.

"I've got to keep on bettering myself," said Nathan, whose first job out of college was as a part-time cashier at a CVS near his home in New Providence, New Jersey. "I'm happy to have a job in this economy. While looking for something better I'm working my way up at CVS. I'm trying to go full-time and then get into their management training program."

Of his training, he said, "I don't know if it makes you smarter. But when you get to each new level on the math and reading tasks, it definitely builds up your self-confidence."

Perched on a couch in the front waiting area of the training center, Nick's mother, Diane, told me that he had been struggling with attention deficit disorder for years.

"During middle school we had every kind of tutor known to man," she said. "Name it, we've done it. This is the first one that's worked." When a brochure from LearningRx showed up in the mail, the scientific aura around the program impressed the Vecchiarellos. They decided to spend $12,000 for a school year's worth of three-times-a-week visits. (Of more than a dozen families I interviewed across the United States about their experiences with LearningRx, that was a typical price, although one spent just $3,000 for a few months' training.) A year later, he was on the honor roll at his high school and off the stimulant medication he had been prescribed by a physician.

"I wanted to be off the pills," Nick said with a smile full of braces.

"Looking at a teacher, listening to them talk for forty-five minutes, used to be a challenge. For the first time now, I'm able to have an interest in what I'm learning. They have a ton of games and puzzles here that require you to pay attention. After training for so long, with them making it more difficult and more difficult, I could apply it to schoolwork. I'm even better at following the drills in varsity marching band."

"It has been a financial strain," acknowledged Nick's father, Richard, a fifth-grade teacher. But, he added, "I think it's made a change in Nick. His grades are better. If it gives him a leg up on life, you can't put a price on that."

The training also seemed to pay off in Nathan's case. The CVS pharmacy where he worked promoted him to assistant manager and, for the first time in his life, he had a girlfriend. Despite those gains, his mother told me she was disappointed that he had given up attending the training classes but said that his decision to stop going could not be blamed on LearningRx.

Searching online, I found a handful of complaints—not all that many, really, considering the number of franchises—posted by parents who said they had paid substantial fees but seen minimal benefit for their children. A handful of people claiming to be current or former trainers for Learning Rx have also posted complaints online about the company's sales tactics, and even manipulation of some customers' test results. But the families I spoke to all had positive stories, many of them insisting that traditional tutoring had been of little benefit but that the exercises practiced at LearningRx, as odd as they seemed, had made a major difference.

"In the first couple of weeks, I really didn't understand what the point of some of the exercises was," said Priyanka Bhatia, a sixteen-year-old junior at Hanover Park High School in New Jersey. In one of the exercises, her trainer at LearningRx had her touch her fingers to her thumb, going from pinky to pointer on her left hand, and the other direction on her right. In another, she was challenged to make up stories with blocks and pencils, incorporating a list of characters and feelings.

"That wasn't one of my favorites," she said. "I wasn't too expressive. But it made me come out of my comfort zone."

After six months of training that included plenty of math- and reading-oriented exercises as well, she found a surprising number of benefits. From all the finger exercises, she said, "My dexterity got a lot better. I play the flute, and my fingers are able to play notes faster now."

On the academic side, she said, "My memorization skills shot up, and my grades got a lot better. I see a noticeable difference. We're doing British literature this year, and I'm just able to understand things a lot faster than I would normally."

The unconventionality of some of the exercises reflects the company's origins. Whereas the founders of Posit, Cogmed, and Lumosity all have advanced degrees in psychology and neuroscience, the founder of LearningRx obtained his PhD in pediatric optometry.

"Largely my focus was on visual training," said Ken Gibson in a telephone interview from his home in Colorado Springs (where, strangely enough, a bear happened to be wandering outside on his lawn while he spoke). "I wanted to train children's eyes to move well and to process what they saw."

Treating children with disorders involving focusing or eye movement, he developed an interest in dyslexia and other learning disorders. "I realized I could help those who had eyes crossed, but I wasn't helping very much with their academic performance," he said. "I started reading the literature about training abilities of every skill, not just visual, but auditory and memory and speed of processing."

By training people to practice the tests that had been developed to measure attention, logic, working memory, visual processing, and other components, he found that many could significantly improve. "I started with the assumption that I could train them all," he said. "But certain skills are harder to train than others."

He shrugged off criticism that he lacks the academic credentials to have developed a training program that aims to improve something as fundamental and forbidding as intelligence.

"This has been a process since 1986," he said. "I don't come from the perspective of an academic. We're not part of Duke University or Harvard. We have to get results to justify the fees that we charge and get referrals."

Much of his confidence, he says, comes from the progress most trainees show on standardized tests.

"We measure every student pre- and posttraining with a version of the Woodcock-Johnson general intelligence test," said Gibson, who began franchising LearningRx centers in 2003. Based on the company's data on over thirty thousand students, he said, "The average gain on a standard IQ test is 15 points after twenty-four weeks of training, and 20 points in less than thirty-two weeks."

An independent study of LearningRx's results offers some support for Gibson's claims. Oliver W. Hill Jr., a professor of psychology at Virginia State University in Petersburg, recently completed a $1 million study financed by the National Science Foundation to test the program's effects. He compared 340 middle-school students who spent two hours a week for a semester at their schools' computer labs using the online version of LearningRx exercises to an equal number of students who received no such training. Those who played the online games, he found, improved significantly not only on measures of cognitive abilities compared to their peers, but also on Virginia's annual Standards of Learning exam. He's now conducting a follow-up study of college students in Texas and said he sees even stronger gains when the training is one-on-one.

My bottom line: LearningRx could be easily dismissed, but it shouldn't be. Yes, it's very expensive, its organization as a franchise is bizarre, and its claims for success are overly dramatic and supported by few peer-reviewed scientific studies. Yet most of the training tasks it offers are based on methods shown to be effective when conducted on computers. And the fact that they are conducted in-person, with enthusiastic trainers following scripts as carefully developed as McDonald's recipes, would appear to be of enormous help to precisely the kinds of people most in need of help. Although Cogmed, Posit, and Lumosity have the studies behind them that LearningRx lacks, I suspect that it would prove at least equally effective. And I will say this: I found myself wishing that a six-year-old girl I know, a family friend who has some attention and self-control issues, could get enrolled in LearningRx, although her family would never be able to afford it. For me personally, though, I scratched it off my list for my training regimen, since I certainly was not about to lay down $3,000 for three months of training when Lumosity's program would cost less than 2 percent as much during the same period.

First-Person Shooter Games

One last form of computerized brain training is commercially available, but it gets no respect from either parents or educators. The discovery that so-called first-person shooter games can actually sharpen cognitive skills dates back to 1998, when a long-haired, math-loving gamer named C. Shawn Green arrived at the University of Rochester. That first year at the university, he became a lab grunt for Daphne Bavelier, a professor of brain and cognitive sciences, helping her out with computer programming and other tasks. In his senior year, Bavelier asked Green to program a test of "useful field of view," the same measure used by Merzenich to test how well people keep track of objects moving on the periphery of their vision. He did the programming, but it seemed to have a bug in it: he and a handful of his friends all performed better than they were supposed to, according to decades of prior research.

"We were doing probably twice as well on this test as the literature said we should do," Green told me. "It wasn't, oh, six percent better. It was much, much better. I kind of scratched my chin and wondered what was going on."

Bavelier tried it; she performed in the standard range. A few other people tried it; they also performed in the standard range. The bug wasn't in the program; the bug was in the brains of Green and his friends. They simply had a freakishly good ability to track objects on the periphery of their visual attention. So Green tried to figure out what made him and his friends so special. A few of them were musicians, but not all. A few were athletes, but not all.

"The one thing that everyone who did exceptionally well at this task had in common," he told me, "was that we all played Team Fortress Classic."

It was just around the time when college campuses across the country were getting networked with high-speed T1 lines. Green and his pals had been plugging in to play the first video game that allowed multiple people to simultaneously play on teams to capture a flag, protect a VIP from assassins, and generally shoot the crap out of other teams.

"I didn't know enough about the scientific literature to understand that this was an unusual thing—you're not supposed to be able to practice one thing and get better at other things," Green said. "So it became my honors thesis."

Together with Bavelier, he conducted four experiments comparing longtime players of first-person shooter games against people who had never played such games, finding each time that regular players were significantly better on tests of visual attention. In a fifth experiment, they trained nine nongamers, including men and women, for an hour per day, for ten days, on Medal of Honor: Allied Assault, a World War II combat game. Compared to a group of eight nongamers whom they trained to play Tetris, a tile-matching puzzle game, those who played Medal of Honor showed marked improvement on the tests of visual attention from their pretraining abilities.

On May 29, 2003, they reported their extraordinary results in the journal *Nature*.

"By forcing players to simultaneously juggle a number of varied tasks (detect new enemies, track existing enemies and avoid getting hurt, among others)," they wrote, "action-video-game playing pushes the limits of three rather different aspects of visual attention. It leads to detectable effects on new tasks and at untrained locations after only 10 days of training. Therefore, although video-game playing may seem to be rather mindless, it is capable of radically altering visual attentional processing."

Together and separately, Green and Bevalier have since published many other studies confirming and extending their observations, as have other groups. Some skeptics question the quality of the hundreds of video-game studies that have piled up, but the evidence has grown so overwhelming that endoscopic surgeons and U.S. military drone pilots now routinely train on first-person shooter games to improve their speed and accuracy. Green has since become an assistant professor of psychology at the University of Wisconsin in Madison, while Bavelier has recently opened her own laboratory at the University of Geneva in Switzerland, a good deal closer than Rochester is to where she grew up, in Paris.

"We want to take these games apart and understand what the entertainment industry discovered without knowing it, that certain video games are powerful for causing brain plasticity and learning," Bavelier told me in an interview from Geneva. "These shooter games are the last kind of activity you would think is good for your mind. But in our hands, they are the most powerful for improving attention."

Not only do the games improve visual attention; Bavelier and Green have found similar effects on auditory attention. Bevalier has even shown that video-game training can improve eyesight, as measured by an individual's ability to perceive subtle differences in shades of gray, something that had previously been demonstrated only with surgery or glasses. Incredibly, that improved ability to perceive shading gradients might even be life-extending: poor contrast sensitivity was found to be among the strongest risk factors for dying in the subsequent nineteen years in a study of 4,097 women who were in their sixties when researchers first met them.

So what about fluid intelligence? She has recently tested the games' effects on standard measures of it but said that because the results have not yet been published in a peer-reviewed journal, I could not quote her account of what the study found. In the meanwhile, though, her published studies already do suggest a benefit.

"It depends a lot on what you mean by intelligence," she said. "There is a whole field that says executive control and the ability to control your attention is a major determinant of intelligence. In that sense, the games are making you smarter. But I'm not telling you that you're going to have a better score on your next exam. I wouldn't claim that. But we keep pushing the envelope, testing new things that might be improved after playing these video games."

What's clear, she said, is that the old dogma that learning never generalizes—the so-called curse of learning specificity, the belief that training can't improve fundamental cognitive abilities—is dead and needs to be laid to rest.

"What Susanne Jaeggi has found with the N-back, what Torkel has shown with working memory training, what others have found with meditation, and what we have shown with video games—they are all different ways of getting at the same underlying mechanisms," she said. "We're all training the flexible allocation of executive and attentional resources. Our brains are constantly bombarded with way more information than we actually use to make decisions and guide behavior. A key aspect is deciding what is task-relevant and what to ignore or suppress as not task-relevant. That's part of everyday life. If you are able to better focus on the relevant features of your environment, without interference from distractions, then you will do better."

Very cool. But without published data showing that playing violent video games would directly increase my fluid intelligence, I was not about to start playing them, particularly since I had never noticed any apparent benefits among my game-playing nephews.

Lumosity, then, along with Jaeggi and Buschkuehl's original version of dual N-back, would suffice for the computerized portion of my own personal regimen. But the quest to build brain power began long before computers were invented, thousands of years ago, at the dawn of human civilization.

CHAPTER 4

Old-School Brain Training

Some of the ancient methods and recommendations for improving the mind have gained new credence as scientists have subjected them to study, while others have not. And some are so basic, and so widely accepted for reasons other than brain building, that they hardly merit discussion here. Do you really need me to tell you that famine, child abuse, drinking alcohol during pregnancy, binge drinking during adolescence, institutionalization in orphanages, repeated blows to the head, and exposure to lead and mercury are bad—bad for the body, bad for the soul, and bad for a young person's cognitive development? Were you really waiting for a new study showing that heavy use of marijuana during the teen years is associated with lowered IQ in adulthood in order to tell your kid to lay off the bong? Is it only for lack of convincing medical evidence that parents would withhold a wholesome diet, a peaceful home, and a good night's sleep from their child?

So let's move on to interventions where the scientific evidence is actually worth knowing. And let's begin with food.

As a pescatarian (a vegetarian who eats fish) and a type 1 diabetic (it developed when I was a skinny eighteen-year-old during college and has required me to take insulin and manage my consumption of carbohydrates in general, and sweets in particular, ever since), I looked for as many well-designed studies as I could find on the relationship between diet and intelligence. You can look, too, by searching on PubMed, the National Institutes of Health's online database of over 22 million medical studies. The problem, you'll find, is that there just haven't been many studies published on the relationship between diet and intelligence, and those that have been published have generally reached negative conclusions.

Don't shoot. I'm just the messenger.

Blueberries, for instance, have been the subject of only a handful of studies in elderly rats, suggesting that supplementation of their diet with the berry slightly improves their memory. If you own an elderly rat, you might consider giving some to the old boy. But the only study I could find in humans involved just nine people and was not randomized, leaving us with essentially no meaningful, reliable information on the merit of blueberries as a means of enhancing cognition. And I say that as someone who owns a blueberry bush and has gone blueberry picking in Maine.

What about the B vitamins: B_6, B_{12}, and B_9 (also known as folic acid)? Although they're all essential nutrients, and many studies have found that people with low levels tend to be at increased risk of developing Alzheimer's, randomized trials in which the B vitamins were added to the diet generally failed to reach the expected conclusion, that they improved memory or thinking skills in those who received them. It's one of those puzzling findings that medical research is full of. As a 2012 review by researchers at Hebrew University in Jerusalem stated: "the majority of these studies have not demonstrated that B-vitamin supplementation has protective or therapeutic cognitive benefit." Another review, published in 2008 by the highly respected

Cochrane Database of Systematic Reviews, concluded: "The small number of studies which have been done provide no consistent evidence either way that folic acid, with or without vitamin B_{12}, has a beneficial effect on cognitive function of unselected healthy or cognitively impaired older people. . . . More studies are needed on this important issue."

As for creatine, a substance produced naturally in the body but sold as one of the most popular supplements taken by athletes, four randomized, blinded trials have been published on its ability to improve cognitive skills: one involving healthy young people, one involving elderly adults, another involving vegetarians, and a fourth involving both vegetarians and those whose diets include meat. The small study involving the elderly found benefits on a variety of cognitive tests after just a week of supplementation. In the other studies, only the vegetarians showed cognitive benefits after taking creatine. That would seem to make sense, because dietary sources of creatine are found primarily in meat, fish, and other animal products, and the level of creatine found in vegetarians' muscles is generally lower than in those who eat meat. But the two studies that included vegetarians were not consistent, with one showing improvement on measures of working memory and fluid intelligence and the other showing improvement only in short-term memory. And neither was large enough to convince scientists who like to see studies involving hundreds of people before they're willing to believe. Personally, I wasn't bowled over by the scanty evidence, and after having my blood level of creatine checked and found to be in the normal range, I decided not to include a creatine supplement in my training regimen.

How about the omega-3 polyunsaturated fatty acids (PUFA) found in fish oils, as a means to prevent cognitive decline and dementia in the elderly? A Cochrane review of omega-3 PUFA supplements, this one published in 2012, identified three well-designed randomized studies involving a total of 4,080 participants, lasting from six months to two years. "The available trials showed no benefit of omega-3 PUFA supplementation on cognitive function in cognitively healthy older people," the Cochrane review concluded. "Omega-3 PUFA supplementation is generally well tolerated with the most commonly reported side-effect being mild gastrointestinal problems. Further studies of longer duration are required. Longer-term studies may identify greater change in cognitive function in study participants which may enhance the ability to detect the possible effects of omega-3 PUFA supplementation in preventing cognitive decline in older people." In other words, three studies involving over four thousand people and lasting up to two years found no benefit—but maybe if they enroll thousands more and study them for ten years, a benefit might emerge. Or might not.

What about fish oil supplements for pregnant women, as a means to optimize fetal brain development? A 2003 study from researchers in Oslo, Norway, published in the journal *Pediatrics*, raised hopes of a positive benefit. They recruited 590 pregnant women to take either cod liver oil or corn oil each day, and to exclusively breastfeed their infants for three months. By that point, only 41 women in the cod liver oil group had followed the instructions and remained in the study, compared to 35 in the corn oil group. At four years of age, mental processing scores were significantly higher in the cod liver oil group. But in 2008, the same group published a follow-up paper in *Pediatrics* reporting no difference in total IQ between the two groups of children by age seven, although a significant difference was seen on a single subtest. Even that modest conclusion, however, was called into question by another study, this one by researchers at the University of Copenhagen. In 2009, they reported in the *Journal of Nutrition* that, well, the title of the study says it best: "Maternal fish oil supplementation during lactation may adversely affect long-term blood pressure, energy intake, and physical activity of 7-year-old boys." The same group reported in 2011 that seven-year-old boys and girls whose pregnant and breastfeeding mothers had taken fish oil rather than olive oil had significantly lower scores on a variety of tests of attention, working memory, and speed of processing. "Early fish oil supplementation may have a negative effect on later cognitive abilities," they concluded in the journal *Lipids*.

Reread that last sentence if you are pregnant, and please think twice about taking fish oil supplements.

The Mediterranean diet—which includes not only fish but plenty of fruits, vegetables, beans, pasta, olive oil, and red wine; not much red meat, sugar, or dairy—has been the subject of dozens of studies, most of which have found that elderly people who follow such a diet tend to have better cognitive abilities than those who do not. But most of those studies have been observational, meaning that scientists ask people what they eat, test their cognitive abilities, and then observe what happens as they age, without randomizing them to follow one diet or another. One recent study did, however, randomly assign one-third of the participants to a Mediterranean diet with added extra virgin olive oil, one-third to the same diet with mixed nuts, and the remaining third to a low-fat diet. Six-and-a-half years later, those on either version of the Mediterranean diet scored slightly but significantly higher on tests of language, working memory, abstract thinking, and other measures of cognitive functioning than those on the low-fat diet, and were slightly less likely to have been diagnosed with either mild cognitive impairment or dementia. But that is just one study, and you don't need me to tell you that changing your entire diet in order to gain some possible future benefit is not easy.

And that is the same problem that researchers would run into if they tried to test the cognitive effects of a diet full of junk food, say, or fast food: it's very, very difficult to get people to agree to change their diet at the request of a scientist and then to follow that diet for weeks or months, let alone years.

There are two foods, however, that have stood up to scientific scrutiny for their ability to increase cognitive performance: one for infants, and one for adults. The first—the first and best food most infants ever consume—is their own mother's milk. Although researchers have long observed an apparent association between breastfeeding and slightly increased IQ scores for the child, some have concluded that the relationship is not causal, and is only due to the fact that mothers who choose to breastfeed are, on average, better educated than mothers who do not. But a careful study published in July 2013 tracked not only whether mothers breastfed their infants, but how long they breastfed. Even after controlling for the mothers' intelligence and socioeconomic background, the study found that each month of breastfeeding resulted in an extra 0.35, or about one-third, of an IQ point on the verbal scale, and about 0.29 points on the nonverbal one at the age of seven. A full year of breastfeeding would be expected to increase a child's IQ by about four points. An editorial accompanying the study noted: "In the United States about 70 percent of women overall initiate breastfeeding, although only 50 percent of African American women do. However, by six months, only 35 percent and 20 percent, respectively, are still breastfeeding."

After weaning, the only food proved to enhance cognition is coffee. It's not just that the caffeine in coffee is a stimulant; a study published in the journal *Neuropharmacology* in January 2013 found that caffeine improves working memory in middle-aged men independent of its stimulating effect. It's not just the caffeine that's beneficial, either; another study, published that same month in the journal *Age*, found that the working memory of elderly rats fed coffee showed significantly more improvements than those fed caffeine alone. And the benefits of coffee last far longer than the couple of hours during which its effects can be felt; a 2012 study in the *Journal of Alzheimer's Disease* found that adults over sixty-five whose blood levels of caffeine indicated they drank coffee regularly were dramatically less likely to progress from mild cognitive impairment to full-blown dementia over the ensuing two to four years. The study, researchers concluded, "provides the first direct evidence that caffeine/coffee intake is associated with a reduced risk of dementia or delayed onset, particularly for those who already have mild cognitive impairment."

These findings, by the way, are in no way influenced by the large cup of coffee I'm sipping right now, as I write on my laptop at a McDonald's on 42nd Street in Manhattan, where I've temporarily escaped from the New York Public Library.

But other than coffee and mother's milk, as passionately as many of us feel about food and its health effects, for now, at least, no solid body of scientific evidence proves that any other dietary intervention makes a difference to intellectual abilities—other than fish oil supplementation for pregnant women, which actually looks harmful for infants' development.

Moving on. How about bilingualism: does learning a second language increase fluid intelligence? A 2010 study from a memory clinic in Montreal, where many people grow up speaking both French and English, found that bilinguals-from-birth were just as likely to develop Alzheimer's at any given age as those who grew up speaking only one language. But immigrants who arrived in Canada speaking their native language, and then had to also master either French or English upon their arrival, tended to develop Alzheimer's almost five years later than others. A number of studies have reached a similar conclusion: learning a second language well enough to be considered "bilingual" does seem to delay the onset of Alzheimer's disease.

That's wonderful. But any and all forms of education are generally associated with a delay in the age at which people develop dementia.

What about actually increasing intelligence, which is the subject of this book and the goal of my training regimen? On that question, the evidence looks mixed at best. A 2009 paper from researchers in Italy did find that seven-month-old infants raised in bilingual homes were better at responding to a computerized test of attention than children raised in monolingual homes. And a series of studies by psychologist Ellen Bialystok of York University in Toronto has found benefits in some, but not all, measures of cognitive control in young bilingual children over those who speak just one language. However, one of the largest studies on the subject, involving 266 young adults, concluded, discouragingly, that "being bilingual does not provide young adults advantages in cognitive processing and that being trilingual results in lower, rather than greater, cognitive control."

Given that the evidence for bilingualism is so muddled, and that mastering a second language—not simply taking a couple years of French or Spanish in middle school, but actually learning one well enough to be considered "bilingual"—takes years of study, it hardly seems fair to put it in the same category as practicing the N-back for four weeks. I'm glad I learned a little Spanish in seventh and eighth grades; it helped when I visited Central America in 1983. And I'm glad I took a month of Italian classes; it helped when I visited my Italian cousins in 1985. But as a means of making myself smarter, I scratched mastering a second language off my list.

So which ancient methods of maximizing brain power *do* stand up to scientific scrutiny? I found three.

Physical Exercise

Mens sana in corpore sano. No cliché has a better pedigree than "a sound mind in a sound body." This fragment of a quotation from the ancient Roman poet Juvenal has been taken for two millennia as an assertion that the two go together—that you can't have one without the other. But is it literally true that physical fitness can make a person smarter—so true that a doctor should prescribe exercise for a person at risk of developing Alzheimer's, a middle-aged professional seeking to regain a youthful edge, or a child struggling at school?

The popular imagination holds two relevant but conflicting stereotypes. On the one hand, we tend to associate fitness these days with intelligence. We like our business executives and politicians trim; Chris Christie, governor of my fine state of New Jersey, was so tired of explaining away his obesity that he underwent gastric bypass surgery in early 2013. But on the other hand, nobody expects hockey players or weight lifters to win any mental competitions. Brawns and brains are not expected to go together. That's why Arnold Schwarzenegger and Sylvester Stallone found their greatest successes in Hollywood portraying big goons.

Research dating back to the 1960s and 1970s found suggestive evidence that physical fitness does affect mental performance. A classic study in 1975, for instance, found that older people who played tennis or racquetball performed significantly better than their non-exercising peers on a variety of simple cognitive tests. Other studies piled up through the 1980s, mostly involving the elderly, but the findings remained only suggestive until two of today's most respected cognitive psychologists dived into a study of swimming.

The actual experiment was carried out by Harold Hawkins, the psychologist I quoted in chapter 1, who now manages the Office of Naval Research's program to investigate cognitive training in hopes of improving the capabilities of military personnel. Jaeggi and Buschkuehl are among the dozen or so teams of researchers he currently funds. Before he joined the ONR, however, Hawkins was on the faculty of the University of Oregon, with a grant from the National Institute on Aging.

"I was talking to Harold one day and he said, 'Art, I collected this data a year ago and I haven't done anything with it. Would you like to look at it?' "

Arthur Kramer, professor of psychology at the University of Illinois at Urbana-Champaign, was named director of the university's Beckman Institute for Advanced Science and Technology in 2010. Twenty years ago, though, he was an up-and-coming cognitive neuroscientist who had never studied the role that physical fitness might play in the brain until having that fateful conversation with Harold Hawkins.

The data that Hawkins had collected proved remarkable. He began by analyzing previous research in the area, which suggested one thing above all: the thinking skills of older adults suffer most when they must divide their attention. To pin down this observation as tightly as possible, he crafted a clever study comparing the cognitive skills of fourteen people between the ages of twenty and thirty-five against an equal number of people between the ages of sixty-five and seventy-four. Sitting in front of a computer screen while wearing earphones, participants were asked to press a button with the middle finger of their right hand if they saw one letter, and another button with the index finger of that hand if they saw a second letter. Likewise, if they heard a particular tone, they were asked to press a third button with the middle finger of their left hand, and to press a fourth button with the index finger of that hand if they heard another tone. First Hawkins tested their speed and accuracy doing only the audio challenge; then he tested them doing only the visual; and finally he combined both audio and visual. Although the speed and accuracy of the older adults were slightly worse than their younger counterparts' when the tests involved only a visual or an auditory component, they fell off a cliff when the two kinds of challenges were combined. Their ability to split their attention had suffered the most due to aging.

To see if this age-related decline could be improved with cardiovascular exercise, Hawkins designed a second experiment involving forty men and women between the ages of sixty-three and eighty-two, none of whom had been involved in a regular exercise program prior to the study. Half agreed to participate in a ten-week aquatics exercise program for forty-five minutes per day at the YMCA in Eugene, Oregon; the other half were asked to keep on doing nothing. At the conclusion of the study, the exercisers were no faster than the nonexercisers in the single audio or visual tests, but they had become significantly faster in the combined audio and visual test. Their ability to multitask had been enhanced in just ten weeks.

"Lo and behold, there were some nice cognitive benefits," said Kramer, who handled the writing and analysis on the paper, published in 1992 in the journal *Psychology and Aging*. "But I was worried. I wanted to know if the work would replicate, if the same kind of thing could be shown in other studies."

Although Hawkins had gotten the exercise ball rolling, it was Kramer who picked it up and ran with it. In 1999, he and nine colleagues at the Beckman Institute, along with one researcher at Bar-Ilan University in Israel, published a study in the journal *Nature*, generally considered the most prominent and respected publication in all of science. Over a period of six months, they reported, 124 previously sedentary adults between the ages of sixty and seventy-five were randomly assigned to either mild aerobic exercise in the form of walking up to an hour a day, three days a week, or anaerobic exercise—stretching and toning.

"In the ageing process," they wrote, using the proper British spelling of "ageing" for a proper British journal, "neural areas and cognitive processes do not degrade uniformly. Executive control processes and the prefrontal and frontal brain regions that support them show large and disproportionate changes with age." It was this kind of age-related decline in executive control that had explained why the 1992 study had found that multitasking suffers most in aging, because it requires not just the simple kind of attention that cats display when watching mice, or dogs display when watching cats, but the rapid, consciously directed shifts in attention—keeping an eye on the clock while watching a *Tom and Jerry* cartoon while studying for an exam—that humans excel at.

So while performance on tasks that didn't require attention shifting was equivalent for the walking and toning groups after six months, the walkers gained significantly over the toning group in task-switching tests. The findings were especially striking because the intervention involved only three hours of walking per week, which improved the walkers' average maximum rate of oxygen consumption by just 5.1 percent.

As expected, given the prominence of the journal and the clarity of the results, the study received a great deal of coverage around the world, from the BBC ("Exercise Boosts Brain Power") to the *New York Times* ("A Good Workout for Older Minds"). Kramer went on to publish over a dozen more studies of exercise and cognitive abilities, including two involving children. In 2010, he published an fMRI study of nine- and ten-year-old children showing that those with greater aerobic capacity had better memory and a larger hippocampus, the seahorse-shaped structure located deep in the brain that is essential to forming both short-term and long-term memories. Another study by Kramer, published in 2012, found that fitter children had higher levels of cognitive control than those who were less fit; remained accurate in their test responses longer; and showed more brain activation of frontal regions during the tests.

One of the few randomized studies of exercise for children was led not by Kramer but by researchers at Furman University in Greenville, South Carolina. Working with African American students in second to eighth grades, they randomized some to participate in forty-five minutes of daily physical education during the 2009–10 school year and compared them to students who did not participate in the program. By May of 2010, the students who participated in the exercise program had improved significantly more than students in the control group on eight of twenty-six cognitive measures (as well as seven of sixteen fitness and body composition measures).

If confirmed in other studies, the findings on children could carry profound implications for school systems, which over the past few decades have significantly cut the amount of time devoted to physical education in favor of basic academics and test preparation. How ironic would it be if, by shorting children on gym and sports, they have undermined the very cognitive abilities they sought to enhance?

In the meanwhile, the effects of physical exercise on the mental agility of graying baby boomers is, at this point, beyond dispute.

"In the past ten years, there have now been at least four meta-analyses looking at published studies," Kramer told me. "They all come to the same conclusion: there is a significant effect of fitness training on cognition. We have lots of studies done all over the world. What's interesting is you're not training any aspect of cognition. You're not learning anything. You're just walking, swimming, or bicycling. It's just three times a week. But despite that, you get better almost across the board in terms of different aspects of memory, perception, and decision making. It's amazing that we find the results we do with such a small change in a person's life."

Kramer's faith in the efficacy of aerobic exercise over other kinds, however, is not universally shared. A 2008 Cochrane review, for instance, questioned whether the cognitive benefits associated with cardiovascular exercise are due solely to improvements in cardiovascular fitness rather than to increased muscle strength and other effects of exercise.

The leading proponent of strength training for improving cognitive function is a spritely forty-year-old jogger, dog lover, and mother of three who holds an impressive list of academic credentials: associate professor in the department of physical therapy at the University of British Columbia; Canada Research Chair in Physical Activity, Mobility, and Cognitive Neuroscience; research director of the Vancouver General Hospital Falls Prevention Clinic; and director of the university's Aging, Mobility, and Cognitive Neuroscience Laboratory.

"My first degree was in physical therapy," Teresa Liu-Ambrose told me. "I practiced for two years, treating mostly athletes, national to amateur. I started working more with an older adult population and went back to get my PhD in 2004, focused on fall prevention and bone health. I did an exercise trial, looking at seventy-five- to eighty-five-year-old women. I noticed they were quite easily flooded in terms of information overload—'I can't come to class today because I need to figure out my tax return,' even though it was January.

"There were quite a few 'aha' moments," Liu-Ambrose continued. "Early in the study, I had to pick up some people to drive them to class, because they were so reluctant or they found it difficult to process how to use mass transit. Near the end of the study, though, everyone was getting themselves there either by buddying up or using transit. One person had been an accountant before retirement. Halfway through the study, she decided she was going to go back and become a self-employed consultant. I thought the transformation was quite significant. That's how I got onto the track of exercise and brain health."

In 2010, she and colleagues in Vancouver published the results of a study involving 155 women between the ages of sixty-five and seventy-five who were randomly assigned to either once-weekly resistance training, twice-weekly resistance training, or (as a control group) twice-weekly balance and tone training. On a standard measure of cognitive control, both resistance training groups improved by over 10 percent, compared to a decline of half a percent in the tone and balance group.

A follow-up study published in 2012 was even more impressive. Liu-Ambrose and colleagues recruited eighty-six elderly women with the kind of minor memory complaints typical of mild cognitive impairment and randomly assigned them to six months of either toning and balance training, resistance training, or aerobic exercise. The aerobics group improved significantly on measures of balance, mobility, and cardiovascular capacity—but not on any of the cognitive measures. In contrast, the group assigned to resistance training improved on tests of attention, conflict resolution, and memory. On fMRI tests, as well, only the resistance trainers showed signs of increased activity in three regions of the cortex.

One reason Liu-Ambrose was glad to see the cognitive benefits from resistance training as opposed to aerobic training, she said, is that many sedentary older adults are unable to engage in aerobics. "A large proportion really don't have that capacity," she told me. Also, resistance training can prevent falls and the kinds of bone injuries that can often prove disabling.

"A key point of our resistance training is it's progressive," Liu-Ambrose said. "And it has to be individualized. We assess people's ability to lift certain weights appropriately. Then we apply the overloading principle. That means you want to work the muscle just a little beyond its comfort level, while still being able to maintain proper form. And we monitor every session for any improvement. If they can do two sets of eight, we increase the load by 10 to 20 percent. We think that progression is important in terms of how resistance training may promote brain health."

Her words resonated: the success of resistance training, Liu-

Ambrose was telling me, depends on the same method that Jaeggi and Buschkuehl use in N-back training. It has to be tuned precisely to each person's initial capacity and then progress as capacity expands. One physical, the other mental, but both entail pushing continuously to the limit, the benefits accruing as those limits slowly expand.

Her regimen, incorporating just six basic exercises—including leg press, hamstring curl, lat pulldown, and seated row—can be seen in two entertaining and beautifully edited videos on YouTube by searching on either Liu-Ambrose's name or "Exercise Is Power."

Despite her findings on the benefits of resistance training, she is quick to endorse aerobics if that is what floats your boat. "If someone is jogging," she said, "certainly from Art Kramer's work, it has consistently been shown to provide benefits."

The real problem, Liu-Ambrose said, is convincing people to exercise.

"We have to approach five hundred people to get one hundred who agree to participate," she said. "And out of that one hundred, only eighty to eighty-five will complete. And we work pretty hard to retain them. The human interaction can make the biggest difference. The instructor plays a huge role in relating to them, showing a genuine interest in their life."

The next biggest challenge, she said, is finding resistance training classes in the community. "For resistance training, there aren't a lot of classes that are widely available. In Vancouver we have some at the local Y, but if you go outside a major metropolitan area, there's not much. Most communities do offer some health classes for older adults, but usually it's light: balance or light aerobics. Even if they use weights, it's not progressive. We definitely need to do more as a society to offer these programs that really help."

As things stand, alas, it would seem that the worldwide decline in fitness, consistent with ever-rising rates of obesity, is literally making us dumber.

Not, however, for members of the U.S. Supreme Court, who have their own justices-only gym on-site behind those famous white marble columns. Justice Ruth Bader Ginsburg, who turned eighty on March 15, 2013, began working out with a personal trainer in 1999, after undergoing treatment for colon cancer. Her hour-long sessions include a warm-up, stretching, weight training, and balancing.

"When I started, I looked like a survivor of Auschwitz," Ginsburg told the *Washington Post* in an interview. "Now I'm up to 20 pushups." Of her longtime trainer, Bryant Johnson, she said, "I attribute my well-being to our meetings twice a week. It's essential."

Case closed. I added physical exercise to my list.

Music

It can't be a coincidence: the two psychologists whose studies became milestones in the science of music as a means of enhancing cognition both began their careers as musicians.

The first was Frances H. Rauscher, who studied the cello before obtaining a doctorate in experimental psychology at Columbia University. In 1992 she joined the Center for Neurobiology of Learning and Memory at the University of California in Irvine, where she and two colleagues conducted an experiment that, for a while anyway, became about as famous as Benjamin Franklin's flying of a kite in a thunderstorm. You almost certainly have heard of the big take-away from the study, known as the Mozart effect: the notion that babies would get smarter, even while in the womb, if their parents played Mozart for them.

But here's what the study actually involved: Thirty-six college students (not babies) spent ten minutes each listening to either silence, taped instructions for relaxing, or Mozart's Sonata for Two Pianos in D Major. Immediately after each listening session, they took a "spatial reasoning" test, the kind that requires a person to accurately imagine rotating a three-dimensional object depicted on a piece of paper. Their average spatial IQ score after ten minutes of silence was 110, and after the relaxation tape it was 111. But after just ten minutes of listening to Mozart, their average score was 119. "Thus, the IQs of subjects participating in the music condition were 8–9 points above their IQ scores in the other two conditions," Rauscher and colleagues reported in the October 14, 1993, issue of *Nature*.

Cue the trumpets and the big bass drums.

Even though the small study involved college students, and even though the paper made clear that the effect lasted only for ten to fifteen minutes, somehow our popular culture concluded that Mozart makes babies smarter, prodded along by an over-the-top book, *The Mozart Effect*, and its sequel, *The Mozart Effect for Children*. By 1998, Zell Miller, then the governor of Georgia, was proposing that his state spend $105,000 a year to provide every child born there with a recording of classical music. "No one questions that listening to music at a very early age affects the spatial-temporal reasoning that underlies math and engineering and even chess," he told legislators. After playing them a recording of Beethoven, he asked, "Now, don't you feel smarter already?" A *New York Times* article quoted the reaction of state legislator Homer M. DeLoach: "I asked about the possibility of including some Charlie Daniels or something like that, but they said they thought the classical music has a greater positive impact."

Two years later, on June 16, 2000, President Bill Clinton, musician Billy Joel, and Viacom CEO Sumner Redstone visited Public School 96 in East Harlem, New York, to celebrate the awarding of

$5 million worth of musical instruments to city schools by the VH1 Save the Music Foundation. Recalling how he learned to play the saxophone during high school, Clinton said in a speech that day: "I might not have been President if it hadn't been for school music."

By then, the journal *Nature* had already published two devastating follow-ups to Rauscher's paper: a meta-analysis of twenty other studies that had tried to replicate her study and found, on average, a gain of just 1.4 IQ points after listening to Mozart, and a new attempt at replication, which found nothing at all. "The results show that listening to the Mozart sonata produced no differential improvements in spatial reasoning in any experiment," the study concluded. "A requiem may therefore be in order."

"The Mozart effect? That's just crap," is how E. Glenn Schellenberg, a psychologist at the University of Toronto, summed up the academic consensus in 2010 for the *Los Angeles Times*. Which is kind of ironic, because Schellenberg is the other leading figure in the study of music-to-make-you-smarter. His reputation, however, has held up far better than Rauscher's.

Like Rauscher, Schellenberg started out as a musician. In 1977 he became the keyboard player for an influential Toronto rock band, the Dishes, which had some local radio hits and once opened for the Talking Heads. In the late 1980s and early 1990s he composed music for three films, including *Urinal* ("a mystery man brings together a group of dead, gay artists to investigate a police response to the dilemma of wash-room sex in Toronto") and *Zero Patience*, a film about AIDS that drew a nice review in the *New York Times* for Schellenberg's "bouncy stylistic hybrid of Gilbert and Sullivan, Ringo Starr, the Kinks and the Pet Shop Boys."

And he threw all that away for a career in academia. Well, nobody ever awarded tenure to the Pet Shop Boys.

Having majored in psychology at college, Schellenberg decided to pursue a doctorate in the field and by 2004 had become a professor at the University of Toronto. That year, he published a study that has since been cited by 363 subsequent papers: "Music Lessons Enhance IQ." Unlike Rauscher's study, which involved a mere ten minutes of listening to Mozart, Schellenberg set out to actually give music lessons to young children for an entire school year and see if it raised their full-scale IQ more than acting lessons or no lessons at all. He recruited 144 children, six years of age, and assigned one-fourth to keyboard lessons, one-fourth to voice lessons, one-fourth to drama lessons, and one-fourth to no lessons. After thirty-six weeks, the IQ of all four groups had gone up slightly—a normal result of entering elementary school—but the music-trained students went up significantly higher than the others. Scores went up by 3.9 points for those who received no lessons, compared to 5.1 points for the students trained in drama, 6.1 points for those trained on keyboard, and 7.6 points for those given voice lessons. "Compared to children in the control groups, children in the music groups exhibited greater increases in full-scale IQ," Schellenberg concluded. The higher IQ scores, he noted, were accompanied by higher grades for the music-trained students at the end of their school year. (Why the voice-trained students did better than the keyboard-trained was unknown.)

As with Rauscher's study, the reaction to Schellenberg's modest results was dramatic.

"Academically the field has gone crazy," he told me. "I'm kind of a grump in this field. I'm really the only person that showed convincingly that formal music training makes you smarter. But I take my own findings with a grain of salt, and I think other people should, too."

As universities have opened departments like the Music and Neuroimaging Laboratory at Harvard Medical School, where studies have poured out showing that musicians have higher-functioning brains, Schellenberg remains leery of their pretty pictures, pointing out that none has involved the kind of randomized experimental design that his 2004 study followed. "I'm not a neuroscientist," he said. "I don't even like neuroscience. But I know enough about science to say that you can't infer causation if your study doesn't follow a proper experimental design. A lot of neuroscientists don't seem to understand that."

It took Schellenberg himself to replicate his own findings, in a 2011 paper on which he collaborated with Sylvain Moreno and five other Toronto-area researchers. The team, led by Moreno, enrolled forty-eight preschoolers to participate in one of two kinds of computerized training: either visual art, emphasizing concepts such as shape, dimension, and perspective, or music, including rhythm, pitch, and melody. After participating for two hours a day, five days a week, for four weeks, only the children in the music group showed gains in verbal intelligence, with 90 percent of the music-trained students showing those gains.

Additional evidence for a cognitive benefit from music lessons comes from London, where a program called the Bridge Project has enrolled hundreds of students from two schools in Lambeth, a working-class neighborhood. A study commissioned by the program compared the math and literacy grades of children participating in the program to those in a control group. The grades of children in the music program, the study found, improved 10 to 18 percent more than those not in it.

These days, Schellenberg told me, he sees the relationship between music lessons and intelligence as a two-way street. "Nature and nurture are almost impossible to separate in the case of music lessons," he said. "I think that smarter kids are more likely to take music lessons, and to stick with them longer, which in turn expands their cognitive functioning even further."

It's safe to say, then, that while listening to a recording of Mozart is not going to make anyone smarter, learning to play an instrument might. The evidence for music training is modest—not as strong as it is for physical exercise or for working-memory games like the N-back. But no study has ever been published contradicting Schellenberg's findings. And at least, at the end of the day, one learns an enjoyable skill, to make music, unlike with N-back, which would be a pointless waste of time if it did not increase fluid intelligence.

I added music training to my list.

Mindfulness Meditation

As the only academic in the world to go by the title "director of contemplative neuroscience," Amishi Jha follows a remarkably hectic schedule. In January 2013, the associate professor of psychology at the University of Miami gave a talk about mindfulness meditation at the World Economic Forum in Davos, Switzerland. On February 6, she spoke at the New York Academy of Sciences on the "science of mindfulness" and the following week led a day-long meditation retreat with Tussi Kluge, the fourth wife and widow of billionaire John Kluge. The month after that, in March, Jha's article about the subject was the cover story of the magazine *Scientific American Mind*. Meanwhile she administers a $1.7 million grant from the U.S. military to study the effects of mindfulness training on soldiers' resilience at Schofield Barracks, an army installation in Honolulu, Hawaii. She has even shared a stage at the Aspen Brain Forum with actress Goldie Hawn, who is funding a program for schools called MindUP.

But don't let the trendiness fool you. A multitude of studies suggest that the ancient practice of mindfulness meditation actually holds promise as a way to improve cognitive abilities, to increase attention, expand working memory, and raise fluid intelligence. Some of the best are by one of the most respected psychologists in the United States, Michael Posner, professor emeritus at the University of Oregon and former chair of its psychology department. The author of hundreds of scientific papers, Posner was awarded the National Medal of Science by President Barack Obama in a White House ceremony on October 10, 2009.

"I was completely surprised by the results of these papers," Posner told me during one of our many conversations. "Most studies assumed it would take months or years to see effects. We saw changes in the white matter of the brain after two weeks. We've also seen substantial changes in behavior and in cognitive control, the attention network."

After publishing a study in 2005 building on Torkel Klingberg's studies to show that a mere five days of computerized attention training can improve fluid intelligence in children between the ages of four and six, Posner was approached by Yi-Yuan Tang, a psychologist and neuroscientist who held positions at both the University of Oregon and Dalian University of Technology's Laboratory for Body and Mind in Dalian, China.

"I wasn't a practitioner of meditation," Posner said. "I've always been a fairly calm person. But Yi-Yuan came to see me and said that he had done work in China on meditation and that he wanted to establish this as a method in a way that would be respectable from a Western scientific point of view. I'd been studying attention for many years, and since he thought he could show changes after just five days, it seemed practical. I agreed to help him design the experiments."

Tang had developed a specific kind of mindfulness meditation that he calls Integrative Body-Mind Training, or IBMT. As described in their first study, published in the *Proceedings of the National Academy of Sciences* on October 23, 2007:

> The method stresses no effort to control thoughts, but instead a state of restful alertness that allows a high degree of awareness of body, breathing, and external instructions from a compact disc. It stresses a balanced state of relaxation while focusing attention. Thought control is achieved gradually through posture and relaxation, body-mind harmony, and balance with the help of the coach rather than by making the trainee attempt an internal struggle to control thoughts in accordance with instructions. Training in this method is followed by 5 days of group practice, during which a coach answers questions and observes facial and body cues to identify those people who are struggling with the method.

Others have described mindfulness meditation as emphasizing moment-to-moment awareness of whatever thoughts, feelings, or bodily sensations come to mind, without judgment or reflection, letting them come and go like passing clouds, while only attention and alertness remain.

For the study, Tang enrolled eighty Chinese undergraduates at Dalian University, half of whom were randomized to attend an initial training session followed by five days of IBMT for twenty minutes per day, while the other half received relaxation training. Before and after, the students were tested on the Raven's standard progressive matrices; on a measure of attention developed by Posner, called the Attention Network Test; as well as for anxiety, depression, anger, fatigue, and the amount of stress-related cortisol in their saliva. On the attention test, those who meditated showed significantly better cognitive control than students in the control test. Raven's scores went up more for the meditators than for controls; and all the other measures improved significantly.

Posner and Tang followed this up with a series of hard-core neuroscience studies to see what was going on in the brain to cause such rapid changes in attention, mood, and fluid intelligence. In 2010 they published another study in the *Proceedings of the National Academy of Sciences*, this one showing that eleven hours of IBMT resulted in greater integrity and efficiency of white matter—the neural wires and cables connecting neurons—emanating from the anterior cingulate cortex, or ACC. Shaped like an upside-down Nike swish, and located a couple inches above and a few inches behind the eyebrows, the ACC is closely connected to the prefrontal cortex. It's known to work hardest during tasks requiring cognitive control and when mental effort is required to learn or solve problems.

Then, in 2012, came their magnum opus, a third study in the *Proceedings of the National Academy of Sciences*, this one looking more closely at the nature of those white-matter changes in the ACC. Among sixty-eight undergraduates at Dalian University, those who practiced Tang's version of mindfulness for five hours over two weeks showed increased growth of nerve fibers in the ACC but not of the myelin sheath that covers each of those fibers like insulation covers electrical wires. Among forty-eight students at the University of Oregon, however, those who underwent eleven hours of training over the course of four weeks showed increases in both nerve fibers *and* myelination. So it appears that the wires laid down in the first two weeks get their insulating coating in the second two weeks.

"Almost everything in neuroscience is disputed by someone, but I don't think these synaptic changes following training are disputed anymore," Posner told me. "We believe that by changing the white matter itself, these connections are improving the efficiency that underlies the behavioral changes."

Does that mean that mindfulness meditation literally makes people smarter? "Nothing is going to work for everyone," he said. "Not everyone will benefit from mindfulness training. In some people it will have bigger effects than others. But I don't doubt that different kinds of training can improve attention, working memory, and intelligence. The basic data is pretty strong."

I added mindfulness meditation to my regimen, my list of interventions now nicely balanced between the age-old methods of physical exercise, music training, and meditation and the modern, computerized approaches of Lumosity and the dual N-back. But there remained two broad, futuristic approaches to cognitive improvement to explore: brain-zapping devices and mind-expanding drugs. To get my bearings, I set off for—where else?—New Orleans.

CHAPTER 5

Smart Pills and Thinking Caps

Startled awake by the ringing of a telephone while lying in a bed in the French Quarter, a block from Bourbon Street, I found myself momentarily forgetting where I was and why I was there. Too many meetings, too many scientists. Reaching for the phone, I remembered: New Orleans. The twenty-second annual Neuropharmacology Conference—this year's theme, "Cognitive Enhancers." All the top "smart-pill" researchers would be there.

"This is your wakeup call," said the automated voice on the phone.

I hung it up, lay back down, and shut my eyes. For just a moment. And then I dreamed that my head, the top of it, was expanding, like bread in an oven. My forehead grew to about twice its normal size. Just as it cracked open with a hideous noise, I saw my brains begin to bubble out.

That's when I woke back up for the meeting.

•—•

"Why do we need cognitive enhancers?"

The second speaker of the morning was Gary S. Lynch, one of the field's longtime leaders. He'd been among the smart-pill pioneers who, in the 1980s and 1990s, on the basis of early studies in mice and other animals, joined or started companies with futuristic sci-fi names like Cortex Pharmaceuticals and Memory Pharmaceuticals, none of which went anywhere. A lot of money had been spent, a lot of magazine articles had been written about "Viagra for the brain" and that kind of thing, but Lynch never quit his day job as a researcher at the University of California, Irvine. He'd even coauthored a book titled *Big Brain*. Which, given my nightmare that morning, I found rather disturbing.

"Every four years," he said, answering the question he had posed at the beginning of his talk, "this blight descends upon America called presidential elections. Anyone following it this year can see the need for cognitive enhancers."

Surrounded by over 250 scientists—so many that the planners of the meeting had switched it to a larger room at the Hilton New Orleans Riverside to accommodate the overflow—I had come to find out if they were any closer to developing drugs that could safely enhance cognitive skills (the key word being "safely") than they were two decades ago, when claims were first being made that success was right around the corner.

In a debate held in connection with the meeting, Nora Volkow, director of the National Institute on Drug Abuse, expressed grave worries about the drugs already on the market. Stimulants like Adderall, which are supposed to be prescribed only for attention-deficit/hyperactivity disorder, are now used by 8 percent of U.S. high school seniors for nonmedical purposes each year, she said. A newer drug, Provigil, approved only for the treatment of narcolepsy, has also come into widespread use among students and businesspeople seeking to gain an edge. But when David Nutt, president of the British Neuroscience Association, asserted that few of those users experience adverse effects due to the drugs, Volkow shot back: "Yes, these stimulant drugs have been used by the military for more than fifty years, but what's interesting is that the British stopped using [them] and it had to do with the fact that use of these stimulants can make a person paranoid. In the United States the use of stimulant drugs has been associated with friendly fire [due] to distortions of perceptions as well as paranoid thinking."

Perhaps the biggest danger of stimulants like Adderall, she said, is that they can be addictive. Even Provigil has been found to be habit forming. Still, Volkow said, "If you can develop a medication that has no adverse effects and you can improve cognitive abilities either by enhancing your attention or your memory, my perspective is, why not? It would be very, very exciting that there could be such medications, [but] again I state *without side effects*, because the current medications that are available do have side effects."

Even if currently available drugs carried no side effects, one question that students and others who use them never seem to entertain is whether they actually work—whether they truly make people smarter, better able to solve problems and remember things, or if they just keep them awake and working longer. To answer that question, Martha Farah, a psychologist at the University of Pennsylvania, presented one of the first studies ever to test thinking and learning skills in healthy, rested young people given Adderall. "The results did not reveal enhancement of any cognitive abilities," she concluded. A trend toward a slight improvement among the lowest-performing students was seen on word recall and the Raven's progressive matrices—but it was not large enough to be statistically significant. Despite the lack of clear benefit, Farah's most striking finding was that participants nevertheless *believed* their performance was better when taking the drug than when they received the placebo. How scary is that? A pill that simply makes college students cockier.

"I want to be really clear that I'm not scolding researchers," Farah told the group. "I'm not trying to be a nihilist or say that there's nothing to this cognitive enhancement stuff. But the extent of our lack of knowledge is really extraordinary."

Following her remarks, in a remarkable breach of protocol, no applause came from the audience. Farah walked from the podium in silence.

A more welcoming reception was given to Barbara Sahakian, who co-organized the meeting and who, despite holding prominent positions at both Cambridge and Oxford universities and speaking in a British accent, actually grew up in Boston. Sahakian presented encouraging results on a study she had done of thinking skills on Provigil or placebo in healthy, rested adults—one of the first such studies ever carried out. Improvements on the drug were seen, her study concluded, on "spatial working memory, planning and decision making at the most difficult levels, as well as visual pattern recognition memory following delay." But on a test of creativity, she reported, the effects "were inconsistent and did not reach statistical significance."

In conclusion, Sahakian said, "We need novel pharmacological treatments." She also endorsed the use of cognitive training and other nondrug approaches to improving cognitive function. "The best effects," she said, "might be if we combine all these treatments."

Most of the rest of the meeting was taken up by presentations on the many dozens of new drugs now in development at universities and drug companies around the world, with names like ZIP, crebinostat, THPP-1, cytotoxic necrotizing factor 1, LSN2463359, and LSN2814617. Some were being tested against Alzheimer's, some to restore clear thinking in people with schizophrenia or depression, and some aimed at enhancing cognition in healthy adults. Most of the new drugs had been tested so far only in mice, however, and some only in test tubes. And the presentations just kept coming.

"Fortunately we can dissolve UBP7089 in a medium which includes calcium."

"The persistent activity of PKM zeta mediates some but not all LTM."

"The LTP we've been enhancing is LTP 1. We had no idea of the existence of an LTP 2 lurking out there."

I started looking around at my fellow attendees. Tweed jackets and bow ties, the kind still worn by scientists in movies, were nowhere to be found. To my left sat some beefy guy wearing a studded-leather wrist band. Next to him was a twenty-something with a shaved head and a camouflage-style shirt. Two rows over sat an overweight guy wearing flower-patterned Bermuda shorts and a polo shirt. A young woman beside me was puttering with her smartphone, while the guy next to her was sipping something from a tiny red bottle. I did a double take and realized it was 5-Hour Energy, the supplement containing caffeine and other ingredients, such as folic acid and vitamin B_6. Seemed kind of funny that a scientist at this meeting would take such a thing. Then I saw another little bottle of it in front of the woman next to me.

And another little bottle of it was in front of me. The maker of 5-Hour Energy had somehow succeeded in placing its product at a meeting of cognitive enhancement. My attention lagging, I opened it up and drank it down. It tasted horrible. But fifteen minutes later, I did feel more alert.

And then a break was announced. Everyone started walking, and I followed, not sure where we were going. Down hallways, up stairs, we walked for over five minutes until we reached a room containing coffee and sugar cookies. Outside, on a patio, a few scientists smoked cigarettes. Between the nicotine, the caffeine, and the sugar, these researchers would stop at nothing in their quest to discover cognitive enhancers.

The most encouraging presentation of the day, oddly enough, had nothing to do with drugs. Tracey Shors, a notably cheerful psychologist from Rutgers University with long, straight blond hair, described a study she had carried out in mice. Although researchers have known for years that new neurons develop continuously in the hippocampus, a section crucial for forming new memories, little has been known about how to augment their survival. Most end up dying. So even though exercise, sex, and Prozac all help new neurons to be born, the question Shors wanted to address was how to keep them alive longer—to become a functional, permanent part of the brain. The secret, she found, was that the mice had to learn new tricks. In her experiments, she challenged them to learn how to stay on top of a rod spinning under their feet—the mouse equivalent of log rolling—while perched over a container of water.

Mice hate water.

So, highly motivated, the mice learned the footwork necessary to stay atop the rotating rod even as it spun faster. The better they got at it, the more new neurons survived in their brains.

"Tasks that are difficult to learn are the most effective," Shors said. "If they simply exercised, they didn't retain the neurons. Learning must occur, and it must require some effort. So if you exercise, you will produce more neurons. If you do mental training you'll keep alive more cells that you produced. And if you do both, now you have the best of both worlds: you're making more cells and keeping more alive. The effort part is key. We need to learn things that are new, and we need to keep it challenging."

What's more, the cycle of neuron survival multiplies over time, she found, their numbers growing like, well, a family of well-fed mice. "Animals that learn something new at the beginning of an experiment," she said, "several weeks later they learn a second task even faster. So brain cells that weren't even born at the time of the first training are now more likely to survive by virtue of the fact that the older cells stayed alive. I guess the message is that if you keep learning throughout your lifetime, as more and more brain cells survive, eventually your brain would explode."

She was joking, right? I think she was joking.

Before the meeting ended, I wanted to know how long it might be before one of the new drugs being developed might actually get on the market. None of the researchers offered estimates during their presentations, so I sat down at the end of the day with one of the silver-haired lions of the field, Tim Tully. Like Gary Lynch, Tully had founded a company back in the 1990s with the goal of putting a smart pill on the market and had seen those hopes dashed. Unlike Lynch, however, in 2007 he left his day job in academia—head of neurogenetics at Cold Spring Harbor Laboratory—to become chief science officer of a new smart-pill company, Dart NeuroScience.

"I quit because when Ken Dart offered me the job, I said, 'That's really cool, but you have to realize it's going to take you twenty years before you make a profit,'" Tully told me. "I thought he'd flinch, but instead he said, 'Yeah, that's about what I'm expecting.' So then I was in, because I had an investor, an owner, who understood that it would take a long time and a lot of shots on goal until we'll be successful. We scientists who once said we'd have a drug in five years, we were naïve. I now doubt that I will necessarily live long enough to see a really safe and effective cognitive enhancer in my lifetime. But I have no doubt that it will come."

Gary Lynch, when we sat down for an interview, expressed a similar mix of caution and optimism.

"The mechanisms of this next generation of drugs being developed are well understood," he said. "We're seeing a lot of beautiful successes in animals. I think we're at the point where we can see they will work in humans, if we can get past the safety bumps. But that's a big if. The safety margins needed for cognitive enhancement are going to be very large. People are going to take these drugs repeatedly—nobody's going to take them just once. So the winner is going to be the first one that comes through with a clean safety profile. The history of neuropsychiatry is that somewhere between the phase 2 study of a hundred people and the phase 3 study of three hundred people the beast jumps out from behind the bush: the drug turns out to have some unexpected side effect. It's going to come down to whether or not they run into the troll under the bridge."

As for drugs already on the market, he said he shared the perspective of most at the meeting: neither Adderall nor Provigil does much more than keep people awake and working. They might help a student finish writing a paper, but they won't help him or her write it better. But then he mentioned another cognitive enhancer that he does consider useful, despite its nasty reputation: nicotine.

"Nicotine has a legitimate enhancing effect," he said. "I self-medicate with it when I'm having a difficult time writing something. I'll sit there chewing a cigar when I hit a tough spot. People say to me, 'Do you ever take any of these drugs you're studying?' Nicotine, I find, helps me out."

Back home in New Jersey, I read through dozens of human and animal studies published over the past five years showing that nicotine—freed of its noxious host, tobacco, and delivered instead by chewing gum or transdermal patch—may prove to be a weirdly, improbably effective cognitive enhancer and treatment for relieving or preventing a variety of neurological disorders, including Parkinson's, mild cognitive impairment, ADHD, Tourette's, and schizophrenia. Plus it has long been associated with weight loss. With few known safety risks.

Nicotine? Yes, nicotine.

In fact—and this is where the irony gets mad deep—the one purpose for which nicotine patches have proven futile is the very same one for which they are approved by the Food and Drug Administration, sold by pharmacies over the counter, bought by consumers, and covered by many state Medicaid programs: quitting smoking. In January 2012, a six-year follow-up study of 787 adults who had recently quit smoking found that those who used nicotine replacement therapy in the form of a patch, gum, inhaler, or nasal spray had the same long-term relapse rate as those who did not use the products. Heavy smokers who tried to quit without the benefit of counseling were actually twice as likely to relapse if they used a nicotine replacement product.

"I understand that smoking is bad," said Maryka Quik, director of the Neurodegenerative Diseases Program at SRI International, a nonprofit research institute based in California's Silicon Valley. "My father died of lung cancer. I totally get it."

Yet for years Quik has endured the skepticism and downright hostility of many of her fellow neuroscientists as she has published some three dozen studies revealing the actions of nicotine within the mammalian brain.

"The whole problem with nicotine is that it happens to be found in cigarettes," she told me. "People can't disassociate the two in their mind, nicotine and smoking. It's not the general public that annoys me, it's the scientists. When I tell them about the studies, they should say, 'Wow.' But they say, 'Oh well, that might be true, but I don't see the point.' It's not even ignorance. It's their preconceived ideas and inflexibility."

I met Quik at the annual meeting of the Society for Neuroscience held in Washington, D.C. Amid thousands of studies presented in a cavernous exhibition hall, the title of hers jumped out: "Nicotine Reduces L-dopa-Induced Dyskinesias by Acting at β2 Nicotinic Receptors."

"A huge literature says that smoking protects against Parkinson's," she said. "It started as a chance observation, which is frequently the most interesting kind."

The first hint of nicotine's possible benefits, I learned, came from a study published in 1966 by Harold Kahn, an epidemiologist at the National Institutes of Health. Using health-insurance data on 293,658 veterans who had served in the U.S. military between 1917 and 1940, he found the kinds of associations between smoking and mortality that even by the mid-1960s had become well known. At any given age, cigarette smokers were eleven times more likely to have died of lung cancer as were nonsmokers and twelve times more likely to have died of emphysema. Cancers of the mouth, pharynx, esophagus, larynx—blah, blah, blah. But amid the lineup of usual suspects, one oddball jumped out: Parkinson's disease. Strangely enough, death due to the neurodegenerative disorder, marked by loss of dopamine-producing neurons in the midbrain, occurred at least three times more often in nonsmokers than in smokers.

What was it about tobacco that ravages the heart, lungs, teeth, and skin but somehow guards against a disease of the brain? Over the course of the 1970s, neuroscientists like Quik learned that the nicotine molecule fits into receptors for the neurotransmitter acetylcholine like a key into a lock. By managing to slip through doors marked "Acetylcholine Only," nicotine revealed a special family of acetylcholine receptors hitherto unknown.

And what a family. Nicotinic receptors turn out to have the extraordinary capacity to moderate other families of receptors, quieting or amplifying their functioning. According to psychopharmacologist Paul Newhouse, director of the Center for Cognitive Medicine at Vanderbilt University School of Medicine in Nashville, "Nicotinic receptors in the brain appear to work by regulating other receptor systems. If you're sleepy, nicotine tends to make you more alert. If you're anxious, it tends to calm you."

The primary neurotransmitter that nicotine nudges is dopamine, which plays an important role in modulating attention, reward-seeking behaviors, drug addictions, and movement. And therein lies the answer to the mystery of why nicotine could prevent a movement disorder like Parkinson's disease, due to its effects on dopamine.

To put the drug to the test, Quik treated rhesus monkeys with Parkinson's with nicotine. After eight weeks, she reported in a landmark 2007 paper in the *Annals of Neurology*, the monkeys had half as many tremors and tics. Even more remarkably, in monkeys already receiving L-dopa, the standard drug for Parkinson's, nicotine reduced their dyskinesias by an additional one-third. Studies of nicotine in humans with Parkinson's are now under way, supported by the Michael J. Fox Foundation.

Other research suggests the drug may protect against the early stages of Alzheimer's disease. A study involving sixty-seven people with mild cognitive impairment, in which memory is slightly impaired but decision-making and other cognitive abilities remain within normal levels, found "significant nicotine-associated improvements in attention, memory, and psychomotor speed," with excellent safety and tolerability.

"What we saw was consistent with prior studies showing that nicotinic stimulation in the short run can improve memory, attention, and speed," said Newhouse, who led the study.

As Newhouse sees it, "Obviously the results of small studies often aren't replicated in larger studies, but at least nicotine certainly looks safe. And we've seen absolutely no withdrawal symptoms. There doesn't seem to be any abuse liability whatsoever in taking nicotine by patch in nonsmokers. That's reassuring."

That's not reassuring: it's totally bizarre. Nicotine has routinely been described in news accounts as among the most addictive substances known. As the *New York Times Magazine* famously put it in

1987, "nicotine is as addictive as heroin, cocaine or amphetamines, and for most people more addictive than alcohol."

But that's just wrong. *Tobacco* may well be as addictive as heroin, crack, alcohol, and Cherry Garcia combined into one giant crazy sundae. But as laboratory scientists know, getting mice or other animals hooked on nicotine all by its lonesome is dauntingly difficult. As a 2007 paper in the journal *Neuropharmacology* put it, "Tobacco use has one of the highest rates of addiction of any abused drug. Paradoxically, in animal models, nicotine appears to be a weak reinforcer."

That same study, like many others, found that other ingredients in tobacco smoke are necessary to amp up nicotine's addictiveness. Those other chemical ingredients—things like acetaldehyde, anabasine, nornicotine, anatabine, cotinine, and myosmine—help to keep people hooked on tobacco. On its own, nicotine isn't enough.

But what about nicotine as a cognitive enhancer for people without Alzheimer's, Parkinson's, or any other brain disease?

"To my knowledge, nicotine is the most reliable cognitive enhancer that we currently have, bizarrely," said Jennifer Rusted, professor of experimental psychology at Sussex University in Britain when we spoke. "The cognitive-enhancing effects of nicotine in a normal population are more robust than you get with any other agent. With Provigil, for instance, the evidence for cognitive benefits is nowhere near as strong as it is for nicotine."

In the past six years, researchers from Spain, Germany, Switzerland, and Denmark—not to mention Paul Newhouse in Vermont—have published over a dozen studies showing that in animals and humans alike, nicotine administration temporarily improves visual attention and working memory. In Britain, Rusted has published a series of studies showing that nicotine increases something called prospective memory, the ability to remember and implement a prior intention. When your mother asks you to pick up a jar of pickles while you're at the grocery store, she's saddling you with a prospective memory challenge.

"We've demonstrated that you can get an effect from nicotine on prospective memory," Rusted said. "It's a small effect, maybe a 15 percent improvement. It's not something that's going to have a massive impact in a healthy young individual. But we think it's doing it by allowing you to redeploy your attention more rapidly, switching from an ongoing task to the target. It's a matter of cognitive control, shutting out irrelevant stimuli and improving your attention on what's relevant."

Of course, all the physicians and neuroscientists I interviewed were unanimous in discouraging people from using a nicotine patch for anything other than its FDA-approved purpose, as an aid to quit smoking (even though studies find it doesn't work for that purpose), until large studies involving hundreds of people establish the true range of benefits and risks. But with so many studies showing that it's safe, and so many suggesting it might well be the most effective cognitive enhancer now on the market, I decided to ignore not only their advice but the advice of my personal physician.

I added a nicotine patch to my list.

•—•

Thumb-tacked to a cubicle wall outside the office of Roy Hoshi Hamilton, director of the University of Pennsylvania's Laboratory for Cognition and Neural Stimulation (LCNS), is a photo-manipulated advertisement for Dos Equis. It shows the suave, gray-haired, bearded actor who portrays the Most Interesting Man in the World sitting next to his usual bottle of beer. Over the Dos Equis label, however, another label had been substituted, showing a brain with bolts of electricity surrounding it. The tagline of the campaign has also been altered. "I don't always non-invasively stimulate my brain," it states, "but when I do . . . it's at the LCNS. Stimulate responsibly, my friends."

It turns out there is a safer and more reliable way to stimulate your brain than with beer, whiskey, coffee, Adderall, Provigil, or any other known drug, whether approved or in development. With surprisingly little public attention, given its stunning successes in randomized clinical trials, the field of noninvasive brain stimulation has produced results that seem straight out of science fiction. Using voltages so low they can be generated with the same 9-volt batteries that power a flashlight, transcranial direct-current stimulation, or tDCS, causes neither convulsions nor almost any other known side effect beyond a mild tingling when current is applied to the skull. The procedure takes just twenty minutes, repeated for a mere five to ten days. Yet studies published in leading scientific journals since 2005 have shown that tDCS can improve outcomes for a variety of purposes that include (deep breath) depression, stroke, traumatic brain injury, long-term memory, math calculations, reading ability, complex verbal thought, planning, visual memory, the ability to categorize, the capacity for insight, and the solving of an inherently difficult problem.

Too good to be true, right? How can a minuscule amount of electricity, applied to the skull for twenty minutes with the same kinds of positive and negative cables used to jump a car battery, make people think better?

When I posed that question to Hamilton in his office on the fifth floor of the university's Goddard Laboratories building, he leaned back in his chair, smiled, and answered, in the classical style of an ancient philosopher, with his own question.

"What is a thought? A thought is what happens when some pattern of firing of neurons has happened in your brain. So if you have a technology that makes it ever so slightly easier for lots and lots of these neurons, these fundamental building blocks of cognition, to be active, to do their thing, then it doesn't seem so far-fetched that such a technology, be it ever so humble, would have an effect on cognition. How could it affect it enduringly? There's this mantra in neuroscience, coined by Donald Hebb: Neurons that fire together wire together. So I have this tool that makes it more likely your neurons will fire. Now, while I'm applying the current I'm going to have you engage in some behavior, a working-memory task, say, or reading aloud some words even though you have aphasia following a stroke, which is my area of interest. So now that network of neurons is being activated in an environment that slightly nudges it, makes it slightly easier for the neurons to fire and the behaviors to be successfully carried out. Then it's not too far-fetched that, when that happens over and over again, during two weeks of practice, those pathways will be reinforced. I agree it does seem very simple. It's not like we're inserting some super-high-tech nanorobots into your brain to clear up your carotid arteries. But it is in accord with our thinking about how brains work. And it does appear to have wide-ranging effects."

With a brown goatee, curly salt-and-pepper hair, and tawny skin that could pass for Italian or Hispanic, Hamilton is actually Japanese-African-American. "Hoshi, my middle name, is my mother's maiden name," he told me. "She's from the Fukushima area. And my father's from Oakland. They met in Los Angeles. My mother came to the United States in the hopes of advancing her English studies. That plan was altered. She and my father got married. I grew up in Long Beach, approximately twenty-seven miles south of L.A."

After graduating magna cum laude from Harvard with a BA in psychology, Hamilton planned to become a psychiatrist. But in his first year at Harvard Medical School, he explained, "I went to a lecture by the behavioral neurologist Alvaro Pascual-Leone. He had just arrived at Harvard as a new assistant professor. He was talking about how blind individuals could perform these remarkable tactile tasks, touch-related tasks, and when they did so, they activated parts of their brain that you and I would use for vision. They're activating their visual cortex when they're touching stuff. Alvaro demonstrated that he could manipulate that by placing a powerful magnet to the back of the blind person's head, zapping that region of the brain. In that instant, their ability to perform those remarkable tactile tasks is reduced. He demonstrated that you can use a piece of technology to manipulate brain activity in a focal, precise way. I thought, 'That's the coolest thing I've ever seen in my life and that's what I'm going to do.' "

That was in the mid-1990s, when Pascual-Leone was pioneering the use of transcranial magnetic stimulation to explore and manipulate the function of brain regions. Hamilton ended up taking three years off from his medical school education to work in Pascual-Leone's laboratory before finally obtaining his Harvard medical degree, cum laude, and then completing his residency in neurology at the University of Pennsylvania, where he is now an assistant professor. Hamilton has now published ten studies in a span of three years demonstrating that tDCS can help adults with dyslexia to read better and can improve the ability of stroke patients suffering from aphasia to find the words they're searching for.

But all those studies lasted only ten sessions. Three miles to the northwest of Hamilton's lab, on the campus of Temple University, neuroscientist Ingrid Olson was conducting a study giving tDCS for thirty days in a row to college students.

"Short-term, it's super safe," she told me. "There have never been any serious adverse effects reported. But we don't know the consequences of getting it for thirty days in a row. It's possible you could make yourself really good at working-memory tasks but that, in doing so, you might get worse at something else."

One reason she and other researchers have been interested in tDCS is that the devices are portable and relatively inexpensive, costing as little as a few hundred dollars. Physicist Allan Snyder of the University of Sydney in Australia has even referred to the devices as a kind of "thinking cap." But the low cost—and the fact that similar devices, approved by the FDA for treating muscle pain, are already on the market—is causing worries for Olson and her fellow sober-minded academics. Videos have already begun to appear on YouTube showing young men experimenting on their own brains with homemade tDCS devices, looking for all the world like they're competing to join the cast of *Jackass*. When applied by scientists like Olson, the positive and negative cables are carefully placed on the skull over the particular brain areas they seek to modify.

"If you put a stimulator on the back of your head," she said, "you could stimulate your brain stem. You don't want to mess with your brain stem."

Olson has already published studies showing that tDCS given for ten days improves memory for the names of people whose fame is past its peak—actors like Barbara Eden, of the 1960s television show *I Dream of Jeannie*, and politicians like Tony Blair, the former prime minister of England. She has also shown that it can improve verbal working memory. When training on a working-memory task for ten sessions over two weeks, the participants who received active tDCS improved twice as much as those who received a placebo version, in which the electricity is turned on initially but then slowly turned off.

"We all have limited time and stamina," she told me. "If you could get the same gains in half the time, wouldn't you want that? Everybody would."

When I said I was hoping to get treated with tDCS as part of my training regimen, Olson said she would need to go through a great deal of bureaucratic wrangling to get her university to allow her to treat me. "We have lots of restrictions," she said. But, she added, "I would not worry about having a relative of mine do it. My husband told me last night that if we keep finding benefits, he wants to be in one of our studies, because he feels like his memory is getting worse."

I explained that I wanted to add tDCS to a regimen including physical exercise, N-back training, meditation, and other interventions.

"That's kind of neat," she said. "We talked in my lab about doing a study like that. We said, why don't we just do this megastudy, where we combine every intervention we know of to improve cognitive abilities, and see what the additive effects would be."

What did she think the effects would be?

"The fact is, nobody knows," Olson said. "If anyone tells you they know, they're bullshitting you."

I added tDCS to my list.

And that was the last of the seven activities and treatments for which I found credible scientific evidence that they could increase fluid intelligence, along with N-back, Lumosity, physical exercise, mindfulness meditation, learning a musical instrument, and wearing a nicotine patch. But now I had something else to figure out: how to actually put these into practice in my daily life.

CHAPTER 6

Boot Camp for My Brain

Improving one's cognitive abilities, I discovered, remains very much a do-it-yourself job. You can't just head over to the drive-through to pick up an order of supersized brains.

Two of the activities I had selected to include in my training regimen, N-back and Lumosity, seemed simple enough. Martin Buschkuehl had already installed a program for doing the dual N-back onto my computer, and Lumosity involved nothing more than going online, typing in a credit card number, and following the instructions. But how to start a physical exercise routine, practice meditation, or learn an instrument? Just figuring this stuff out was a cognitive challenge in itself.

Since college, I had claimed to be a regular jogger, and it was mostly true into my thirties. But somewhere in my forties the three-miles-in-thirty-minutes-three-times-a-week routine had devolved into two-miles-in-thirty-minutes-whenever-I-get-around-to-it. Simultaneously, the 165 pounds I had once carried on my five-foot-nine frame had increased by about a pound a year, so that I now stood on the precipice of 200 pounds, which I considered appalling, but which I had been slipping toward like an asteroid falling into a black hole. Along the way I had joined and dropped out of various gyms, joined a local softball team for a couple of summers (and played somewhat better than anyone else on the team ever gave me credit for), bought two or three bicycles to watch them rust, bought a stationary exercise bicycle that wobbled when I pedaled, and entered a 10K race in which I came in almost dead last, just ahead of somebody's obese grandmother.

My problem is I hate routines. I'm Mr. Spontaneity. If you ever saw *Ocean's Eleven*, you might recall the scene where Matt Damon tells Brad Pitt that the casino owner, played by Andy Garcia, is a "machine," because every day he arrives at the Bellagio at precisely 2:00 p.m., works at his office until 7:00 p.m. "on the nose," goes onto the casino floor, spends three minutes talking to his manager, greets the high rollers, then leaves at 7:30 to meet Julia Roberts for dinner. Well, that's not me. I never wake up at the same time, I don't eat meals at the same time, I have never met Julia Roberts, and I follow an exceedingly warped internal clock that makes sense to me and nobody else. And it worked fine for me for many years, thank you very much, but it wasn't working now.

So I looked into my soul and faced the inevitable: I would have to join the insane over-the-top "boot camp" exercise class, Accelerate Performance Fitness, that my wife, Alice, had been attending for the past four years, run by our next-door neighbor, Patsy Manning.

Picture this petite, cheerful woman whom you mistakenly take for average until you see her biceps. Wow! Alice and I both adore Patsy and her fellow exercise-fanatic husband, Andy. They are two of the friendliest, funniest, most upbeat people we know. They have two adorable kids, who are seemingly always outside on their front lawn, throwing softballs and kicking soccer balls and playing games and laughing and having fun. I have always admired and held the whole family in awe, but I knew I could never be like them. They are into exercise the way I've been into writing and reading since sixth grade. What you have to understand about Patsy is that she leads her boot-camp exercise class every Monday, Wednesday, and Friday at 5:15 a.m. and again at 9:30 a.m., then on many days she heads out to continue her own exercise program with a boxing trainer, then she also plays soccer, plus she trains a couple times a year for a "mud run" and an "urbanathlon," where she and Andy run obstacle courses that would exhaust a Navy SEAL, and they look good doing it.

One time I did stop by the park to watch Alice participate in Patsy's boot camp, and it looked way worse than I had imagined: running and hopping and skipping and shuffling up and down stadium stairs for a half hour, then lifting weights, doing sit-ups, and racing around an athletic field for another half hour. Why would anyone do these things if a North Korean dictator wasn't forcing them to?

But now I needed to join the cult. If I was really going to exercise three times a week just as the participants in the studies had done, combining the cardiovascular training that Arthur Kramer advocates with the muscle-building resistance training that Teresa Liu-Ambrose advocates, Patsy's boot camp was the way to go.

So late one morning in September 2012, Patsy had me meet her at Montclair's Brookdale Park to teach me the routines one-on-one that I would be doing in the class. First she had me do an "easy" run around the quarter-mile track. The rest of this "easy" warm-up included jumping jacks, high-knee running in place, and some lunges and stretches. I was sweaty and breathing hard, but we had only just begun. She led me to the old cement stadium steps overlooking the track.

"Usually we'll begin with a couple of easy runs up and down the steps," Patsy told me, "then three runs all-out, up and down the stairs as fast as you can go. Then we do side shuffles up and down, first facing left, then facing right. Then we do fast-feet: you run up two steps, down one step, up two steps, down one, up two, until you reach the top. You want to give it a try?"

"Not really," I said, "but I'll do it anyway."

A single run up and down the stairs left me wheezing and sweating, but I then tried the side shuffle and the fast-feet.

"You're doing great," said Patsy.

"I am not," I said.

"Don't be so hard on yourself," she said. "You're looking good."

"You are out of your mind."

Next, placing my hands against the heavy iron railing at the bottom of the steps and my feet on the second step, I did fifteen angled push-ups. Then, turning around with my hands behind me on the railing, I did fifteen dips (face forward, butt to the ground). Finally she had me hang from the railing with my legs and arms loosely wrapped around it, and then pull up tightly fifteen times, like a monkey playing on a tree branch.

"Okay, let's go onto the field."

Patsy set out two orange cones, about forty yards apart, and had me run back and forth between them. First it was a normal run, then it was a backward run. Then I had to skip. (The last time I had skipped, when I was about nine, it seemed a lot easier.) Then I had to leap, leap, leap across the field.

"Oh, my God," I said, leaning over, my hands on my knees, my lungs heaving for breath. "I am really having second thoughts about this."

"Why?"

"Seriously, do you think I'm ready for this?" I asked her.

"Sure, you're doing fine."

She reminded me of the scene in *Groundhog Day*, where Bill Murray tells Andie MacDowell, "Gosh, you're an upbeat lady!"

"But you really are doing good," she insisted. "Everyone has a hard time at the beginning. It's just a process."

Next she showed me how to do a scissoring sideways move called the "karaoke." Your left foot crosses in front of the right; the right foot steps farther to the right; your left foot crosses *behind* your right foot; your right foot steps farther to the right; and so on. I did it slowly, feeling like I was going to trip over my own feet.

And then it occurred to me: I am as ill-suited to these kinds of physical exercises as some people are to cognitive exercises. I felt like a total clod, as out of my league as some people might feel trying to read *Moby-Dick*.

"Did you bring your weights?" Patsy asked.

I had none to bring. So she lent me two of her ten-pound hand weights and proceeded to demonstrate a series of arm lifts from various positions. After another ten minutes, it was time for cool-down stretching exercises. With sweat stinging my eyes, I followed along, barely able to cross my legs in the basic yoga meditation position.

"So what do you expect to get out of this?" Patsy asked when it was all over. "Do you think you're going to raise your IQ?"

"More like my fluid intelligence," I said and gave her the thirty-second version of what that is and how it's distinct from an ordinary IQ score.

"And you're going to train for just three months? Why only three?"

"That's three times longer than most of the training studies have lasted," I said. "Plus I'm combining all kinds of things. I figure if it requires some huge life-changing commitment that goes on for a year or whatever, nobody's going to do that anyway. I want to make it reasonable and realistic."

"But if your fluid intelligence does go up, how will you know which one of all the things you did really made the most difference?"

"I don't see that it really matters. I'm not doing it as a scientific study to determine which part is most effective. It's still such a novel idea that intelligence can be increased, I just want to see if it's possible at all, so I might as well throw the kitchen sink at it. If there's no effect of any of it, that will be incredibly damning."

We had reached our cars in the parking lot. Patsy opened her van and put her bag of equipment into the trunk. And then she asked the question that everyone else, including my editor and my agent, had asked.

"But what if it doesn't work?"

"Well, that would be interesting to know," I said. "Like I said, it's not a scientific study. I'm just one person. But after three months of boot camp, it's pretty much impossible that I wouldn't be stronger and more physically fit, right? So if increasing fluid intelligence really works, it should be equally impossible that three months of all these various exercises won't result in improvement. We'll just have to see."

I had played guitar since my thirteenth birthday, when my father bought me a cheap, used, warped acoustic one. In college I'd formed a punk band, the Mutations. (Our three greatest hits were "I Hate You," "I Want Your Body," and "Electrocutes.") But it was during the summer before my senior year, when I spent six weeks on a work-abroad program pasting labels to the spines of books in a London library, that the seed of a musical fantasy was planted in my brain.

A friend I'd made that summer invited me to spend the weekend at his family's country home. Sometime during the weekend, we visited the nearby home of his uncle. When we walked in, this middle-aged bald guy—my friend's uncle—sat on a wooden chair playing a strangely malformed, stubby version of a classical guitar, its soundboard shaped like a teardrop rather than a woman's body, with a rounded rear, like a watermelon; its sound hole covered with intricate

Medieval-style carving; its tuning pegs located on a piece of wood cocked ninety degrees back from the fret board. It was ancient and mysterious, the music emanating from it unlike anything I'd heard before: deep, haunting, and devastatingly beautiful.

Thus I came to know of the Renaissance lute. As two people can meet at a bar and know at once that they have found the love of their life, so I knew: when I got old and bald like this guy, I was going to get me one of them and play it. Probably I'd have a glass of sherry beside me, an adoring wench, a fire crackling in the hearth, and a couple of old English sheepdogs sleeping at my feet.

This vision hibernated in the depths of my mind for decades, like a cicada, even as the hair on my head made way for its reemergence, when at last I came across those studies by Glenn Schellenberg about musical training and intelligence and knew: my lute's moment had come.

But where the heck do you find a lute for sale and a lute teacher to give lessons? I had never seen another one since that time in the English countryside. They don't sell any at Guitar Center. I went onto craigslist and eBay but found only a few offered for sale in the greater New York area, all of them handmade, none selling for less than $1,800, most priced at $3,000 and up. I was not about to lay out that kind of change, so I started searching for lute instructors, in hopes that I might rent one. A site for the Lute Society listed instructors across Europe, South America, North America, and Japan, but only two in New York City, and a single one in Manhattan. I e-mailed the Manhattan teacher, Michael Calvert. He was from England, had toured Europe and South America giving concerts on both the lute and the classical guitar, and had been teaching for years. After I e-mailed him, I spoke with him by telephone, and he agreed to see if anyone he knew in New York was willing to rent me a lute. Two weeks later, he'd found no one.

Then I found an advertisement on craigslist for a lute for $445, from a woman named Theresa, who lived out in Pennsylvania Dutch country, a couple hours' drive from my home. It all seemed somewhat shady, for an improbably low price, but Michael Calvert said he knew of the Pakistani company that makes the kind she had, that they usually sell for around $900 new, and that they are perfect for a beginner. He looked at the images of the lute that Theresa sent and said it looked fine. So on a rainy afternoon in September, I drove out to Bethel, Pennsylvania, to meet Theresa at a Valero gas station.

Sitting at one of the two tables inside the station's snack section, Theresa turned out to be a refugee from Manhattan's early-music scene, a singer and instrumentalist who had met some guy who was a Mennonite and followed him to Pennsylvania to get married and have children. Now she was separated from him, missing Manhattan, trying to rebuild her life, and hoping to unload the lute that she no longer played. She opened up its case. Knowing absolutely nothing about lutes, it looked gorgeous to me. Plucking at the strings produced the kinds of sounds I recalled from so many years earlier. She threw in an instruction booklet, a set of strings, and a tuner. As in a drug deal, I paid her $445 in cash and headed home.

Awaiting my first lute lesson, I looked into finding someone to instruct me in practicing mindfulness meditation. Back in my twenties, I had actually gone to a Zen temple in Chicago one Sunday afternoon to see what meditating was all about. About ten people sat cross-legged and silent on a cold stone floor, while a guy in a loose black kimono walked slowly among us. After about an hour, I had started nodding when I felt a hard whack on the head. The guy in the kimono stood over me, holding a wooden stick, flat like a ruler. He looked at me sternly and continued walking. This was actually a standard Zen practice, I later learned, at least in some quarters, to keep meditators meditating. But that was it for me. I left.

Now I went online to see what kind of mindfulness meditation classes were offered in my area of New Jersey. Since Montclair is such a liberal bastion, crawling with artists and actors and journalists and filmmakers, I figured it would be easy. But the classes I found were mixed with other New Age pursuits, like yoga or "angels and healing" or lecture series about wellness. I just wanted plain-vanilla mindfulness meditation without any of the mix-ins, the kind that Michael Posner and Yi-Yuan Tang had found to be associated with increases in fluid intelligence. The local YMCA offered a meditation class on Tuesday nights, but it wasn't specifically focused on "mindfulness" meditation.

Finally I decided to just order a guided mindfulness meditation CD created by Jon Kabat-Zinn, a longtime champion of the practice, founder and former director of the Stress Reduction Clinic at the University of Massachusetts Medical School and author of the best-seller *Full Catastrophe Living*. It felt like a cop-out to be settling for a CD in place of a class or personal instructor, but with all the other approaches I was testing, it would have to do.

And so, along with a nicotine patch, I had my regimen figured out. Sitting on a flight heading home from St. Louis, where the fMRI scan of my brain had taken place in October 2012, I took out a pencil and a yellow legal pad and wrote out my training schedule. I would wake up at 6:00 a.m. instead of my usual 8:00 a.m., spend twenty minutes on the N-back, twenty minutes on Lumosity, and twenty minutes meditating. Then Patsy's boot-camp exercise class would take an hour. Then I would shower and drive into the city for my lute lesson. Boot camp was only three days a week, and the lute lesson only once a week. Plus I'd have to spend some time each day practicing the lute at home. Still, most days my actual training time would be just two or three hours. Seemed reasonable. No, it seemed awesome. What an adventure! I was thrilled to finally begin my great brain-training regimen.

•—•

And then the alarm clock went off at 6:00 a.m.

I swatted at the damned thing in the darkness before the October sunrise to turn it off. Two hours later, I woke up and spent my usual hour drinking coffee, reading the paper, and attaining full consciousness. Then it was time to get in the car and drive over to Brookdale Park for my 9:30 boot-camp class. First, though, I ripped open the foil packet containing my seven-milligram nicotine patch. I'd bought a package of fourteen of them at CVS, sold from behind the counter. Even the CVS brand was about $38. I stuck the circular, inch-wide patch to my arm, just below my shoulder, and headed out.

Patsy's acolytes numbered about fifteen. Most looked in their thirties and forties. Only three of us weren't wearing an athletic bra, but that was enough for me. Running around the track for the warm-up, I was among the last to reach the finish line. Doing the jumping jacks, the knee-high running in place, and the lunges, I felt momentarily sick to my stomach, but it passed. On the stairs, I wobbled up and down them about half as many times as the rest of the group. On the field, she had us do a sequence: lie on the ground, do a sit-up, jump up and leap into the air as if tossing a basketball, then back down onto the ground for a sit-up, and repeat for fifteen times. The third time I tried to stand, my head went dizzy and I saw stars. I perched on my knees for a minute to catch my breath.

"Pick up the pace!" Patsy called enthusiastically.

"Kiss my ass," hissed somebody beside me. Eric was the wise guy of the group. Thank God for Eric.

So it went for another half hour. I went home, took a shower, and collapsed on the couch until 1:00 p.m., when it was time to drive into the city for my lute lesson.

Exiting the elevator in the apartment building on West 99th Street where Michael Calvert gave lessons, I heard, coming through a doorway across the hallway, a challenging piano piece being played. Turning the corner to Michael's door, I heard another piano playing in the distance.

Michael opened the door with a sweeping gesture, like a bow. Between his British accent, bushy goatee, and thinning hair hanging over his ears, he seemed to have stepped out of a Shakespearean play. He welcomed me into his living room, dominated by a grand piano. His wife, he explained, is also a music teacher.

"Is everyone in this building a musician?" I joked.

"Actually, the building management has for many years favored musicians," he said. "It's quite extraordinary. I know the practice schedules of all my neighbors. If they don't play it's like the clock has broken."

I opened the lute case and he took a good look at it.

"It looks to be in very nice shape," he said and began tuning it. My lute had seven sets (or "courses," as they're known) of paired strings, plus one more at the highest pitch. Altogether that made fifteen strings, instead of the six on a guitar, each one tuned with a simple wooden peg in a hole, without the usual mechanical tuning keys that all guitars have. As a result, the pegs can easily slip loose from their tuning.

"An old joke from the Renaissance is that if a lute player lived a hundred years, he would have spent fifty of them tuning the instrument," Michael said.

After twenty minutes of tuning, he handed me the instrument. Because of its rounded back, without the slim waist of a guitar, it kept wanting to slip off my leg. He pulled out a piece of the thin spongy foam used as air-conditioner filter and told me to place it on my thigh. The lute nestled into it and stayed put. Then he pointed down to the floor, where he had a metal footstool, and instructed me to place my foot on it. Now the lute felt truly at rest in my lap.

Next he focused on the placement of my hands, for which my experience on the guitar proved to be a handicap. Never much of a folk-style finger picker, I had generally used a pick to strum away at all the strings for rhythm sections or pluck individual ones for solos. Now Michael instructed me to place my right pinky against the face of the lute just below the strings, to steady the rest of my hand, and then to allow the rest of those fingers to hover gracefully directly over the strings. Each of those four fingers, then, would be used to pluck at the strings. It felt alien and awkward, but he insisted that the proper hand positions would be essential to learning the instrument.

My experience with the left hand was even more of a hindrance. I had always straddled the neck of the guitar with the meat of my thumb. But because the fret board of a lute is far wider than that of a guitar, with more than twice as many strings, I was now supposed to just gently touch the neck of the lute with my thumb tip, so that my other four fingers would be free to reach around to the entire fret board. Again it felt incredibly awkward, but I tried my best.

Now he placed onto his music stand the instruction booklet that Theresa had included with the lute, and asked me to play the first exercise, in which each string is played in order, from high to low: third fret, open fret, third fret, open fret, third fret, open fret. Getting each of my ten fingers to follow its own separate orders was impossible, like choreographing ten hungry toddlers to dance in a ballet. Just gaining the sense of where each string was in relation to the others, so that I could move between them with a steady rhythm, would take a lot of time.

The hour up, Michael wrote down the name of a better instruction booklet that I should buy, as well as the name and phone number of a man in Boston who makes lute strings by hand, so that I could order a full new set.

•—•

Back home, I sat down to use the N-back program that Martin Buschkuehl had installed on my laptop. I clicked on it, but the program wouldn't load. Clicked again. Nothing. Went online, searched for N-back, found a site called Soak Your Head (www.soakyourhead.com) offering a free version of the dual N-back, clicked into the game, and began at 2-back.

A woman's voice slowly read aloud twenty letters in a row. Only seven different letters were included in the sequence: *m*, *q*, *f*, *g*, *l*, *r*, and *s*. But they kept coming in random order, and I had to hit the letter L on the keyboard whenever a letter repeated from two times before. At the same time, while watching eight tiles in a three-by-three grid (the center one left empty), I also had to hit the S key whenever the same tile lit up from two times before in the sequence.

After the twenty-item sequence ended, the screen showed the number of 2-back repetitions that I had correctly identified for both the spoken letters and the visual grid: sixteen out of twenty audio and seventeen out of twenty visual. When I scored at least 90 percent accuracy on 2-back, the next sequence would challenge me to keep track of 3-back repetitions. Score less than 80 percent accuracy and it would send me down a level. Between 80 and 90 percent, I stayed at the current level. So it went for twenty times—that is, twenty sequences of twenty items each. On my fourth and twelfth sequences, I scored well enough to get promoted up to 3-back, but then each time scored poorly enough to get demoted back down to 2-back.

Altogether, it had taken just over a half hour. By the time I had checked my e-mail and made a few telephone calls, it was dinnertime. And then Alice and I were supposed to head out to an art exhibition by a fellow member of Patsy's boot camp. Lumosity and meditation would have to wait. But after just a day, I was already impressed by three commonalities these three very different activities shared.

First: I really sucked at all three. And this is a serious thing. While children are routinely forced by their teachers and life in general to do things they're not good at, by middle age many of us have figured out what our strengths and weaknesses are and have learned to stick with what we do best. We've found careers we're good at; we hang out with people we like and who enjoy our company; our hobbies are things we've been doing for years. We are no longer, as Michael Merzenich put it, engaging our mind at the cutting edge of its abilities; we have become users of mastered skills. So, for me anyway, doing N-back, joining Patsy's boot camp, and taking a lute lesson was like jumping into the ocean on New Year's Day. Could doing stuff you suck at be an essential ingredient to getting smarter?

Second: They all felt nearly impossible to do. It wasn't just that I wasn't good at any of them, but that even just sucking was really, really hard. During all three activities, the thought occurred to me that I was in way over my head. My mind kept saying, "This is crazy" and "I can't do this." But I kept doing it anyway. Could mental blood, sweat, and tears be part of the cognitive-growth process?

Third: Each of these activities demanded not only that I pay attention to multiple streams of information simultaneously but that I control multiple responses, whether manipulating each of my fingers on both of my hands while playing the lute, listening to the stream of spoken letters and watching the lit-up tiles during N-back, or moving my left leg here, right foot there in coordinated fashion up and down the chipped cement steps while Eric muttered "go shit in your hat" over Patsy's exuberant cheerleading.

•—•

The next morning, I reached 3-back five times, but again, each time scored so poorly that I was kicked right back to 2-back.

Then I logged on to Lumosity, typed in my credit card information, set up an account, and began playing the five games the screen said would be "today's training session." First was Playing Koi, where I needed to keep track of each individual koi fish swimming in a pond as their numbers grew. Then came Lost in Migration: five birds flew in formation, and I had to press the correct key as quickly as possible to show whether the center bird was pointing left, right, up, or down. Chalkboard Challenge displayed two equations; the goal was to quickly decide whether the left one was bigger, the right was bigger, or the two were equal. For By the Rules, a series of symbols appeared with varying qualities—color, size, shape, and so on—and I had to figure out what the current rule was for including or excluding each symbol; only green symbols were accepted for one sequence, while only those with thick black outlines for another. Finally Face Memory Workout offered up a scenario in which I was taking orders from customers in a restaurant and had to remember their names and what they had ordered.

After finishing, I received my initial scores for speed, problem solving, flexibility, attention, and memory, plus an overall score. Clicking a box, I could see how I compared to other people in my age range. My overall score was in the 43rd percentile, meaning that more than 57 percent of Lumosity's users in my age range scored better than I did. My reaction was a mishmash: embarrassment, annoyance, disbelief, anger at myself, and one other thing. As President George Herbert Walker Bush said when Iraq invaded Kuwait back in 1990: "This shall not stand."

But before I could unleash my wrath and deliver pain and destruction upon my competitors, it was time to meditate.

In our spare bedroom, I popped in the first of three CDs in the instructional set from Jon Kabat-Zinn. What a calm, soothing voice the man has. I followed along as he told me to lie down on my back. I didn't have a yoga mat, and the floor had only a small throw rug from Target, so I just lay down on the hardwood floor. But the room is rather small, so I had to scooch forward to avoid having my head touching the couch or my feet bumping against the music console.

The goal was not intentionally to relax or calm down, Kabat-Zinn's recorded voice stated, but just to pay attention and notice without judgment. Any thoughts or feelings I had were fine, but I was supposed to just notice them, not hang on to them, and let them go. But however I tried to follow the instructions, he emphasized, there was no wrong way. Unlike the other elements of my regimen, mindfulness meditation required no effort. Alertness, yes; effort, no.

Kabat-Zinn then began directing my attention to parts of my body, from the big toe on my left foot all the way to the tip of my head. It was very relaxing. Even though I kept thinking of researchers to call, meetings to attend, and studies to read, I also kept allowing those thoughts to recede. And as they did, as I let go of the struggles against the N-back and Lumosity and all the rest, I entered that curious state of observing my own observations, aware of my own awareness. Taking it one step further, as Kabat-Zinn's voice murmured along, I realized I was not just noticing myself and my own thoughts and feelings and sensations but, in so doing, I was also observing myself observing myself.

How bizarre: 3-back consciousness.

•—•

Over the days and weeks that followed, my performance slowly but steadily improved at boot camp and Lumosity, on N-back and the lute. My Lumosity score quickly jumped into the 60th percentile and continued creeping upward. Boot camp never felt any less awful, but my pace quickened, the nausea and light-headedness receded, and sometimes I could nearly keep up with Sarah, who was four months pregnant, and Cathy, who had just turned sixty. On the lute, I learned to play an early Renaissance piece that took up an entire page of musical notation, even as Michael kept emphasizing that it was more important to practice playing each phrase correctly, with the precise fingering indicated, no matter how slowly, than to rush forward: accuracy above all. It was so unlike the way I fooled around on the guitar, playing songs I'd learned as a teenager. The skill I was practicing now was not just playing the lute, but maintaining my own attention. My mind was the instrument.

On the N-back, within just a few days, 2-back magically became easier and I was soon able to reach 3-back regularly. My competitive instincts flaring, I started shouting to get myself pumped up between the twenty-item sequences. Laugh if you like, but I actually began imitating the voice of a preacher. "Reach up to touch heaven! Reach now, brother Daniel!" And when I finally did reach and remain at 3-back for a couple sequences in a row—then was promoted to 4-back a single time—I sang out, "I have been to the mountaintop!" But progress was in fits and starts. Sometimes my attention wandered, my performance worsened. And by far the biggest obstacle was just getting myself to sit down to do all the things on my list. After years of resisting routines, I couldn't settle into one now, which of course meant that I often ended up reaching eleven at night without having completed my Lumosity or N-back exercises. But when I did, and spent an hour after dinner practicing the lute, it occurred to me: Isn't this what smart people do? Perhaps an organized life goes hand in hand with an organized mind. Then again, if that's true, why have so many brilliant artists, writers, and inventors been famous for living screwball existences?

Whatever the answer, my one failure with the brain-training regimen I had devised was mindfulness meditation. I did it about seven times over the course of a few weeks. I enjoyed its contrast with my other activities and with the rest of my life. It felt pure. I really wanted to keep doing it. The problem was that I had too many interviews to conduct, meetings to attend, papers to read, and people to meet. And I had seriously underestimated the true amount of time involved in all the other activities. Lumosity took nearly a half hour per day; N-back, about forty minutes; lute practice, another forty minutes, plus two and a half hours to drive to and from New York for my weekly lesson; and two full hours every Monday, Wednesday, and Friday morning for boot camp. Not to mention I have a wife and daughter, who sometimes like to see me and who at other times have been known to play the radio or television or knock on my office door to ask me a question or shout up the stairs to tell me that dinner is ready. Evildoers! Mindfulness meditation just seemed to be an activity better suited to a single person, or to one with a much bigger home, so that the noises and demands of a busy life can be set aside. With great reluctance, I kicked it out the door of my moving vehicle and sped on.

But hold on. Pull over. What is this fluid intelligence that I was attempting to increase? Where are its gears and levers, biologically, within our brains? And why, in the fullness of time, did nature give humans so much of it?

CHAPTER 7

Are You Smarter Than a Mouse?

Consider the common bacterium *Escherechia coli*. Consisting of only a single cell, doomed to live and die in the lower intestinal tracts and feces of warm-blooded animals, it has neither brain nor nervous system. Yet nature provides it with the means to carry out intelligent behavior; indeed, it has a kind of memory. It moves toward nutrients and away from toxins, keeping track over an approximately two-second span of whether its local environment is getting more or less hospitable. How can that be?

Intelligent search—the imperative to seek out and explore—turns out to be fundamental not just to *Star Trek* or the human spirit but to most life on Earth. Searching nonrandomly (intelligently, that is) is the simple, great, basic, and profound challenge facing all living things capable of movement. Even tree roots move (slowly!) toward water; creatures of every kind, microscopic and elephantine, move toward food and drink and away from dangers; toward the right temperature, the right amount of humidity and light, the right sexual partner. And what's really mind-blowing is that the means by which creatures search their surroundings is very much like, and the evolutionary basis of, the way we search our minds to remember and reason.

"The basic idea is that the landscape inside your head is not that different from the spatial world around you," began Thomas F. Hills, a psychologist at the University of Warwick in Coventry, England, when I spoke with him via Skype. "We hunt around in our heads the same way flies hunt around for sugar on a table."

Hills, coeditor of a 2012 book, *Cognitive Search: Evolution, Algorithms, and the Brain*, published by MIT Press, had his epiphany about the evolutionary basis of search while conducting basic research (re-search: careful or diligent search) on *Caenorhabditis elegans*, a soil-dwelling nematode, or roundworm.

"I owe it all to that little worm," Hills said. "It turned out that the machinery they use for search is the same you use for search. The worm neurons and the human neurons are engaged in the same fundamental task. It's called the optimal search problem: you've got to know when to explore and when to exploit."

If we lived in a universe in which all resources were uniformly spread out—equal amounts of sugar, water, gold, jobs, good ideas, and good-looking sex partners—then we could search randomly without any strategy and we would all do equally well. You could stay in Boise or you could move to Hollywood and be equally likely to bump into Leonardo DiCaprio. But we definitely do not live in that kind of world. In this universe, resources come in clumps, or patches. Some places are better than others.

The inevitable logical strategy for creatures hunting in a world of patchy resources, where some neighborhoods are rich and some are poor, is called area-restricted search. As Hills has written, area-restricted search refers to "an individual's ability to restrict search to the local area where it has recently found resources before transitioning to more wide-ranging, global exploration."

The single-celled *E. coli* engages in area-restricted search using a strategy called run-and-tumble. As it encounters an increasing gradient of food resources while floating about in your intestines, it moves its tiny flagella, its little whiplike propellers, in a counterclockwise motion. This sends it straight forward in a "run" toward the higher density of food. But when the amount of food begins to drop as the bacterium moves forward—when the level is lower than it was about two seconds earlier—the flagella switch directions, to move clockwise. Because of the shape of the flagella, this causes the *E. coli* cell to "tumble" in a random direction.

For *C. elegans*, area-restricted search means slowing down and pirouetting when it finds something it likes, so that it hunts around in the nearby area as intensively as possible, on the assumption that more of the same will be nearby. When it stops finding goodies, it stops turning, so that its search becomes less intensive and more extensive, on the assumption that anywhere but here has got to be better.

"I spent so many years studying the way this little guy moves around, and thinking about area-restricted search," he told me, "and then it just occurred to me: we're doing area-restricted search in our minds the same way it does area-restricted search in space. To navigate internal space, just like to navigate external space, you have to know when to give up."

I asked him for an example.

"So if I ask you to list all the animals you can think of," he said, "you'll begin by listing pets, like a dog and a cat. Then you'll give up on that category and jump to another, farm animals. You'll list cows and pigs and sheep until you start running out. Then you'll go to Africa: giraffes, lions, apes. Then to the ocean: whales, sharks, tuna. You search an area until it starts running low, then you switch."

In bugs as in brains, however, the real question is when to run and when to tumble: when, that is, to keep obsessing over a single tiny area and when to jump around.

"There are two ways to err," he told me. "You can give up too soon or you can stay too long. People with ADHD, they're just jumping around too quickly between goals. Addicts, they're perseverating on a particular patch. They keep thinking about the cigarette, the cigarette, the cigarette."

Even short of those extremes, Hills said, "These kinds of questions face us every day. I have to decide how long to edit a paper I'm writing before I submit it. I have to decide which research project to pursue and how long to pursue it. It happens even in a department store: how long will I look for a gift for my wife in this store before I try another store? It's just crazy. Area-restricted search is what we do."

Biologically, Hills has found that just as in the human prefrontal cortex, dopamine plays a key role in the ability of *C. elegans* to stay locked on target or drift off. When its dopamine-sensing neurons are destroyed or blocked with drugs, the worm is incapable of performing area-restricted search; it remains on an extensive search even when food is available in a local area. When its ability to sense dopamine is restored, the worm resumes its usual pirouettes to remain in a tight, area-restricted search in the presence of food. And when the worm's neurons are spiked with an extra dose of dopamine, its rate of area-restricted search increases accordingly.

Dopamine signaling has weirdly similar effects in the human brain. For instance, patients with Parkinson's disease, marked by low levels of dopamine, tend to have difficulty concentrating and staying focused. Once they receive the dopamine precursor L-dopa, however, they sometimes become obsessed with sex and gambling.

"The kinds of drugs I give the worms to make them search more or less locally are the same kinds of drugs that doctors give patients to treat various mental disorders," Hills said. "Someone with Parkinson's is having trouble with ideational perseveration. They find themselves wandering around in this space of ideas. So what do you do?

You give them L-dopa. Someone who's a drug addict, on the other hand, has overactive dopamine machinery. If you block that machinery, they lose their addictive tendencies."

In one of his latest findings, Hills published a study of the "animal fluency task," in which people are asked to name as many animals as they can in a minute. Among 185 adults between the ages of twenty-seven and ninety-nine, he found that older people spent less time in each logical category (like jungle animals or farm animals) before switching to another category. As a result, they generated fewer animals in the allotted time. "They jump around in their heads too often," he said. "They give up too quickly."

They tumbled, that is, when they should have run.

In case you were wondering what all this running and tumbling has to do with intelligence, it just so happens that performance on the animal fluency test and on similar tests of quick list making is closely related to working memory and fluid intelligence. A 2013 study in the journal *Memory and Cognition* found that individuals with higher levels of working memory not only list more animals but generate more categories and draw more animals from each category than do those with a lower level of working memory. Fascinatingly, when the researchers prompted participants with a list of possible categories (which they were free to use or ignore), the gap between those with higher or lower working memory shrank significantly. And when the researchers went further, insisting that participants generate their list of animals from the categories provided, the gap disappeared entirely.

So it's not that "smarter" people have "better" long-term memories; it's that they simply generate more categories and search through them more diligently when trying to remember. They are better at exploiting the clumps of sugar on the tabletop of the mind.

"They're more strategic and efficient in how they search their memories," said a leader of the study, Nash Unsworth, a psychologist at the University of Oregon. "They break it down into clusters."

Unsworth reached similar conclusions, but in a way cooler fashion, when he and the same two collaborators—Gregory Spillers, also at the University of Oregon, and Gene Brewer of Arizona State—conducted a study of people's ability to remember their Facebook friends. After testing the working memories of about a hundred students from the University of Georgia, they selected twenty-four who had scored in the top 25 percent and twenty-one who had scored in the bottom 25 percent. They then gave each of them eight minutes to list as many of their Facebook friends as they could. Although the students in both groups each had about the same number of friends listed on Facebook, those with high working memory remembered significantly more than those with low working memory: 81.9 friends, on average, compared to 66.5. And when Unsworth and colleagues asked the participants to explain how they knew each person, those in the high-working-memory group had used more clusters or categories—16.6 clusters compared to 13.8 in the low-working-memory group—and they also recollected more people per cluster.

"High-working-memory people are more likely to quickly search these different contexts," Unsworth told me. "What about my softball team? What about work? What about my dorm? Whereas low-working-memory people just randomly search and hope. It's a strategic aspect of searching. And when I say strategic, I'm not saying it's a trick people are using, like memory tricks to memorize a deck of cards. I think it's a fundamental difference at the root of memory."

Unsworth sees differences between high- and low-working-memory individuals not just in how strategically they search their memories, but in how carefully they make their memories—how they encode them—in the first place. "Where a lot of people fail," he said, "is they don't encode well. They don't place a memory in a context. The real key for remembering is that, whatever strategy you used for encoding, you have to use the same strategy for getting it out."

While such a conscious approach to remembering may seem artificial, it reminds me of another good friend from Beloit College who got a PhD and now teaches history at the University of Wisconsin. Christopher J. Simer has always impressed me and everyone who knows him as brilliant, with an outrageously detailed memory of historic events. While I was talking with him one afternoon about the subject of this book, he revealed to me that when he's trying to master the details of a particular war or historic period, he quite consciously places names, dates, places, and events into their contexts, breaking them down by trunk, bough, and twig. To remember them, he just climbs that same tree.

Not quite all of our cognitive abilities can be explained by studying worms and bacteria. But how did nature get from them to us, evolutionarily speaking? Until talking with Seth Grant, a professor of molecular neuroscience at the University of Edinburgh, Scotland, I had been under the impression that evolution shapes our genome only with the tiniest of chisels, individual mutation by individual mutation. But at least twice in Earth's history, Grant explained, something big has happened: the evolutionary equivalent of the giant asteroid that hit Earth and wiped out the dinosaurs. One such giant step was taken a half billion years ago, when a primitive sea creature was born with the most dramatic genetic mutation imaginable: a duplication of the entire length of its parents' genome.

"There are certain types of mutations that are exceedingly rare, which have occurred only a handful of times in history," Grant told me. "The remarkable thing is that around 550 million years ago, this mutation on a grand scale occurred in a single animal that had two entire copies of its genome, and it survived and it had offspring which also had double the complement of DNA. And its descendants became the vertebrates."

And then it happened again. "This duplication event was followed by a second duplication event," Grant said. "Two times two is four. That made four genomes in this ancestral vertebrate. It gave rise to all the species with advanced intelligence."

The beauty of having four sets of the same genes is that, with only one necessary for the creature to grow and function, the rest are free to slowly, randomly mutate—for nature to play with them, if you will—until, one day, lo and behold, one of those oddly mutated versions of a gene turns out to be useful after all.

"It took another 150 million years for these genes to diversify enough to become useful," Grant said. "By the time animals were crawling out onto the land, these genes had become biologically extremely complex, which is why vertebrates are the beautiful, sophisticated animals they are."

In a pair of studies published in the journal *Nature Neuroscience* in December 2012, Grant and colleagues examined how the gene variants, freed up by the double-doubling of our ancestral genome, came to play an essential role in complex cognition. Testing mice's ability to learn, using a computer touch screen that rewarded them with a food pellet when they poked the correct answer with their noses, Grant examined the unique roles of the four variants of a gene called *Dlg*. All four of the variants code for a structural scaffold that holds the synapses between neurons in place. Without *Dlg1*, the basic progenitor of the variants, Grant found that a mouse embryo will simply not survive; it's essential to life. Without a functioning version of *Dlg4*, simple operant conditioning—learning to increase or decrease a behavior as a result of positive reinforcement or punishment—

will not occur. A functioning version of *Dlg3* is necessary for visual discrimination. And in a scientific coup, Grant tested not only mice without a functioning copy of *Dlg2*, but four people born with a similar mutation, one that arises spontaneously and has been associated with the development of schizophrenia. Using a similar computer-touch-screen test for the people as the one employed with the mice, Grant found they suffered from similar cognitive failings. Without a normal *DLG2* gene, the people "made significantly more errors than healthy control subjects from the general population in tests of visual discrimination acquisition . . . cognitive flexibility . . . visuo-spatial learning and memory." They also showed less accuracy in tests of sustained attention. The results showed, he concluded, that "*Dlg2*'s role in complex learning, flexibility, and attention has been highly conserved over 100 million years."

But fear not: the dignity and grandeur of human intelligence compared to that of a mere mouse has been preserved typographically, if not biologically: in mice, the gene family is indeed spelled *Dlg*, with the last two letters being lowercase (and they don't call 'em "lower" for nothing), but in humans the scientific convention is to spell them all uppercase: *DLG*.

Yeah! Sticking it to the mice!

It's important to note that no single gene—not *DLG* or any other—has yet been shown to directly affect IQ scores by much more than a single percentage point. A widely heralded study published in 2012 in *Nature Genetics*, for instance, found that a variant in the *HMGA2* gene resulted in brain size of about a half percent larger and IQ of about 1.3 percent greater, on average. Although many studies of families, twins, and adoptees suggest that the intelligence of biological parents accounts for about half of their children's intellectual abilities, the early hope of geneticists that a handful of genes could be found to control that variance is now considered to have been naïve. Perhaps genes, like letters of the alphabet, must be combined into the equivalent of words, sentences, and paragraphs to spell out anything as complex as intelligence.

Equally naïve is the assumption that animals' brain size is directly proportional to their intelligence. Neanderthals had larger brains than modern humans, and a sperm whale's brain weighs about eighteen pounds on average, compared to the typical human's brain weight of about three pounds, or about fourteen hundred grams. At the other end of the scale, crows, jays, and other corvids have brains that weigh around ten grams, on average, yet they have been found to be better at problem solving than most mammals, including dogs. Even the humble mouse, with a brain weighing in at less than a half gram, punches outside its weight class.

"Right now the mice are smarter than my two-year-old daughter," said Sheena Josselyn, a senior scientist at the University of Toronto's Hospital for Sick Children. "If mice can live undetected in your kitchen for a year, they're not so dumb. The problem is, mice and rats and other nonhuman animals don't talk. It's really tough for scientists like me to explore their memories. But all you have to do is ask them the right question and they can tell you the answer. We use touch screens that they poke with their nose. They amaze me with what they can do."

Remember the Fox television show *Are You Smarter Than a 5th Grader?* In November 2011, results of the ultimate extreme version of that kind of contest were reported in Seattle at the annual meeting of the Psychonomic Society. Gene Brewer, the Arizona State researcher who collaborated on the Facebook study mentioned above, was the first author of a paper entitled "Working Memory in Rats and Humans." According to the abstract, "The quintessential instrument used by animal researchers for measuring working memory in rodents is the radial-arm maze. We constructed an 11-arm human version of the radial-arm maze and assessed individual differences in maze ability, general-fluid intelligence, and working memory capacity." A key phrase in the abstract's conclusion caught my attention: "Human behavior in the maze paralleled that found in rodents."

I asked Brewer if that meant what I thought it did: had he actually asked people to compete against rats in a maze, almost literally running in a rat race, and found that the people did no better than the rats?

He had. I begged him, then, to please, please let me come out and run the maze, too. He was willing when we spoke, but later his collaborators turned out to be reluctant. They were concerned, I think, that the media might sensationalize this idea of a "man vs. mouse" intelligence competition.

"It is kind of eyebrow raising, I guess," Brewer told me. "But to step out on a limb and then say that because the rats learn this maze as well as humans do, that we have similar cognitive abilities, that's probably not a good idea. We just happened to notice a peculiar similarity between how well our volunteers learned it and how well the rats did. To really answer whether different species have similar types of cognitive repertoires would take much more work. The bugaboo is definitely language and symbolic thought. We can always ask people to give verbal reports of the strategies they use to complete a task, but you can't do that with animals. The thing that's tough about animal cognition, you're studying a thing that you can't detect except through behavior."

Still, Brewer was happy to explain how the study was designed. Radial mazes for rats are generally just a few feet wide, with tunnels extending from a center spoke like the arms of a Ferris wheel. Food is placed at the end of each arm, and the rats are tested to see if they can remember which of the arms they have already gone down, without repeating. For his study, Brewer added a twist to his eleven-arm radial maze to make it especially challenging: the rats would find no food pellets if they ran down each arm in order, the equivalent of going along a clock face from one o'clock to two o'clock to three o'clock. Nor would they find any if they went down every other arm, as if going from one o'clock to three o'clock to five o'clock. Rather, they had to go down every *third* arm of the maze, the equivalent of going from one o'clock to four o'clock to seven o'clock. Hard as that was for a rat to learn, the real trick was to remember which ones they had already gone down after the first circuit.

To replicate the same test in people, Brewer and his colleagues constructed the walls of an eleven-arm radial maze on a basketball court, out of plastic tarps. Center court was where the eleven arms of the maze met in the middle. At the end of each arm was money.

"But it wasn't money they were going to keep," Brewer said. "We've now collected data on 150 participants. Animals need a reward to motivate them, but humans usually find motivation in just trying to succeed at a task. That's a big difference between rats and humans."

Yes, that's one of them. But what was it like for the participants who tried to complete the maze?

"When you're going through that maze, man, you're good for about seven or eight arms," Brewer said. "And then you're, like—what? Where am I? You just lose it. As you start coming around the second time, it's just very hard to keep track of eleven arms when you can't do it in a simple order."

•—•

Although Brewer was reluctant to infer from his study any conclusions about the intelligence of mice, a neuroscientist at Rutgers University in New Jersey had recently come out of the closet on the question.

"I once was reluctant to use the term 'intelligence' in mice, because it's such a loaded term," said Louis D. Matzel when we met at his office in the psychology building on the university's Piscataway campus. "The first grant proposal I submitted to NIH, in 1992, used the word. The reviewer said there is no place in biology for studying intelligence because everybody knows that intelligence is just a social construct. That gives you a sense of why some animal scientists are so reluctant to discuss it. I even had a dean here at Rutgers who once told me I shouldn't discuss intelligence because it's such a difficult topic. But she was an idiot and got fired soon after, so I felt vindicated."

I was beginning to like this guy. Matzel is a wiry middle-aged character with graying brown hair and a trim mustache. He wears Converse high-top sneakers, smokes cigarettes outside his building, and likes to climb mountains in the winter with his teenaged son. Formerly married to fellow Rutgers psychologist Tracey Shors (the one described in chapter 5 who made the quip at the "Cognitive Enhancers" meeting about an exploding head), Matzel had Christmas lights hung over his office windows—in September. "About five years ago I got too lazy to take them down," he said. Lying atop a bookshelf was a CD called *Until We're Dead* by a band I'd never heard of, Star Fucking Hipsters. On the cork wall near his desk was a photograph of the late Sid Vicious, bassist of the Sex Pistols, and his late girlfriend, Nancy Spungen, whom Sid had been suspected of murdering. Beside it was a photograph of Ian Curtis, the late singer of the British neo-punk band Joy Division. "You don't know Joy Division?" he said. "Look them up on YouTube. There's an excellent documentary that came out about them a few years ago."

Not all his photographs were of dead punk rockers. "Right behind you is a picture of a snail," he said. "*Hermissenda* is a pretty little sea snail. But that was a terrible field I was in. I had some ideas about mice around the year 2000, so I stopped working with snails and never looked back."

Matzel's latest studies of mice have offered perhaps the most astonishing replication imaginable of Jaeggi and Buschkuehl's findings. Before he could conduct the studies, though, Matzel first had to design a test from scratch that would accurately measure working memory in mice. Not only did no such test exist, no proof really existed that mice *have* a working memory. Most researchers, like Matzel, assumed that a mouse needed to have a working memory to keep track of things in its environment. But nobody had ever really proved it.

To do so, Matzel used an apparatus so obscure, it had been described only once in a 1981 paper: the dual maze, consisting of two eight-armed radial mazes placed side by side in a small room whose walls are decorated with bold designs. The idea was to allow the mice to explore a few arms of one maze, then place them in the second to explore a few arms of that one, then put them back in the first—and see if they could remember which arms they had already explored in that first one, so that they wouldn't repeat. Like sailors navigating the ocean by the stars, the mice could look up beyond the confines of whichever maze they were currently in to see the walls of the room and thereby orient themselves to the room as a whole—and, thereby, to the other maze. In a small laboratory across from his office, Matzel decorated the western wall with a large black *S* and a dangling string of tiny lights. On the eastern wall he hung a large black plus sign and another string of lights, but with fatter bulbs. He placed a poster of crudely drawn star shapes on the southern wall and an absurd drawing of cartoon characters from an Adult Swim cartoon on the northern wall.

"The thing is," Matzel said, "if I ask the mouse to work on only one maze, he's really good at it. A well-trained animal will typically make no errors. He'll navigate around and get his eight pieces of food. With two mazes, it's really hard. At first they all make many errors. But here's what we found: over days of trying it again and again, they get good at it. A smart mouse will eventually get really good at it and make no errors. This is a task where some rodents are at least as good as humans. So they do have a working memory."

I asked him what he meant by a "smart" mouse.

"We test intelligence in the mouse on a battery of six learning tests, each one different from the other. Occasionally, we get an animal who turns out to be the best on all six tests compared to fifty other mice. That's a really smart mouse, we would say. We teach them to avoid a shock or avoid a bright light. Navigate through a dry maze. Navigate in a water maze. We also have a reasoning task."

The reasoning task requires the mice to make an inference by exclusion. "I show an animal a star symbol," he said, "and it learns to walk over to an object shaped like a circle; underneath is a treat. So the animal learns that when it sees the star, if it goes to the circle, it gets a piece of food. Star means circle. Then I train it on a square and a triangle. If you see a square, you'll get food under the triangle. So square means 'go to triangle.' And then one day I show the animal a symbol it's never seen before, let's say a crescent, and out in the field of objects it can choose from, it might have a triangle, might have the circle, and a novel object it's never seen before. It looks at the triangle and the circle and thinks, 'It can't be under those, so it must be under the novel object.' It infers by exclusion that the food must be under the novel object. This is the amazing thing—mice are decent at doing that. This kind of reasoning task is considered a quintessential example of humans' ability to reason. And yet mice do it. I'm so amazed they can do it. Because I don't believe my dog can do it. My dog just seems really stupid. For years I'd let my dog out in the yard on a rope, and he'd always wrap himself around a tree."

Using the reasoning task, Matzel demonstrated that, just as in humans, a general intelligence factor can be discerned in mice: those who are better at the reasoning tasks tend to be faster at learning the other tasks. Likewise, those who do better at the working-memory task as measured by the dual maze also tend to do better on the reasoning and learning tasks. But where things get really interesting is in Matzel's amazing mouse version of Jaeggi and Buschkuehl's training studies: in 2010, he reported that the animals whose working memory he trained by having them practice on the dual maze actually got smarter on tests of general cognitive abilities. Finally, and most significantly of all, mice who trained on the dual-maze task when they were younger showed less age-related loss of attention and learning abilities by the time they had reached the mouse equivalent of old age. Matzel and colleagues concluded in that study: "These results suggest that general impairments of learning, attention, and cognitive flexibility may be mitigated by a cognitive exercise regimen that requires chronic attentional engagement." Or, as he told me, "It was our intention to manipulate working memory in mice and see if that manipulation had a direct effect on their intelligence. In fact, Jaeggi's work demonstrates the same thing we found."

If working-memory training increases intelligence in mice, imagine what it might have done for Sid Vicious.

CHAPTER 8

Defenders of the Faith

Randy Engle *is* working memory. He was hardly the first to study it; that honor goes to Alan Baddeley, the British psychologist who first proposed a coherent theory of working memory back in 1974. But more than any other psychologist living or dead, Engle—the psychologist quoted at the beginning of chapter 2 speaking so eloquently about the difficulty of measuring either love or intelligence—demonstrated why it is so important and how it connects so closely to fluid intelligence. In a paper published in 1999, since cited by nearly 1,500 subsequent studies, Engle and three colleagues described a series of eleven memory tests they gave to 133 undergraduates at the University of South Carolina. Some were simple tests of short-term memory—remember a list of words or numbers—while others were the kinds of working-memory tasks that require manipulation of the information to be recalled. They also gave two tests of fluid intelligence, one of them the Raven's progressive matrices, and also noted the students' verbal and math SAT scores. Using complex mathematical formulae to see how the multiple tests related to each other, Engle and his colleagues concluded that "working memory shows a strong connection to fluid intelligence, but short-term memory does not." That is, the better a person does on working-memory tests, the smarter he or she tends to be. Both working memory and fluid intelligence, they argued, "reflect the ability to keep a representation active, particularly in the face of interference and distraction."

But what is it about this ability to avoid distraction that relates so closely to fluid intelligence? In another influential paper, written by Engle alone and published in 2002 in the journal *Current Directions in Psychological Science*, he opened with an outrageously absurd example:

> I am an avid baseball fan, and when I am listening to a game on the radio, particularly if the game involves the Atlanta Braves, my wife will occasionally tell me something that she would like for me to do. However, often, and especially in the middle of a tense game, I will not even notice that she is talking to me. Does this ability to block out information have anything to do with working memory? Is there some relationship between the ability to control attention and the amount of information that can be kept temporarily active in memory?

Engle was not, of course, suggesting that his ability to ignore his wife during ball games should be seen as an indication of his intelligence. Or was he? In his own self-mocking but highly revealing way, Engle was getting at something powerful about the nature of intelligence: the ability to zero in on something and let the rest of the world fall away. As obnoxious as it may be for a spouse, or a friend, or a pet waiting to be fed, every artist, inventor, composer, mathematician, writer, entrepreneur, scientist—every person engaged in demanding cognitive work of any kind, not to mention every reader or student or moviegoer seeking to appreciate that work—must turn over their attention, wholly, to the work at hand. In doing so, for better or worse, they must temporarily shut out distractions.

In one of our many telephone conversations, I asked Engle to describe what he considers to be his chief contribution to the field of psychology.

"What I have shown over and over again is the relationship between those two big variables, working memory and fluid intelligence," he told me. "The important thing for both of them is attention control, your ability to focus your attention on me as I'm talking and block out any distraction, so you can willfully and volitionally move your attention from one thing to another. That variation in attention control is important not just in cognition, but also in behavioral and emotional control. It helps people with post-traumatic stress disorder block out intrusive thoughts and memories. It's one of the most important determinants for whether someone will become functional again after a diagnosis of schizophrenia. After that paper in *Current Directions in Psychological Science*—and it remains the most widely cited paper in the history of that journal, I'm proud to say—people in social psychology, people in psychopathology, in schizophrenia research, and these other areas began focusing on my work. It was that paper that drew the attention of people outside of cognitive psychology to what I do."

Engle went so far in the *Current Directions* paper as to state that working-memory capacity may be "isomorphic to general fluid intelligence"—fancy talk for saying they might be the very same thing defined by two sets of tests. Given, then, his role in establishing the primal connection between working memory and fluid intelligence, thereby paving the way for the studies of training working memory that followed from Klingberg, Jaeggi, Buschkuehl, and so many others, it is more than a little surprising—or, perhaps at a level that students of Greek mythology could appreciate, perfectly predictable—that he is now universally regarded as the leading opponent of their work, a fierce and vitriolic critic, Grand Inquisitor, and chief defender of the faith that working-memory training does not increase fluid intelligence.

He has told me, at one time or another during conversations by telephone and in person, that Jaeggi and Buschkuehl's studies are "egregious," that he "shudder[s]" at the "absurdity" of their claims, which he has compared to claims made back in the 1980s, since disproved, that "cold fusion" could be achieved with a desktop device. Of their breakthrough paper, published in 2008, he has accused them of selectively publishing only the data that supports their views. "They cherry-picked," he said. "That comes very, very close to an American Psychological Association ethical violation that would get you kicked out of the APA. That paper is as close as it comes. There's no good evidence for this stuff now. Yet they continue to do the same thing. That one paper sent hundreds of smart people off on a wild-goose chase, in my opinion." Asked about Michael Merzenich and Posit Science, he told me, "If you listen to Merzenich, he can fix everything from appendicitis to xenophobia. There's a lot of the snake oil salesman in Merzenich. He started out as a legitimate scientist working with mice. I'm sure he was a good rat scientist." Of Torkel Klingberg and Cogmed, Engle said: "That's a whole new category of sleaze. The commerce is driving these claims in a very, very big way. Cogmed, if you google it, it looks like Cogmed can be used to solve everything from arthritis to lumbago. Pearson is making a fortune off this, and the researchers are in on the cult. Cogmed in my opinion is an absolute sham." He predicted that studies would soon "dismantle" the false claims. Many others shared his view, he said, but added, "I'm the first one to tell you this, because I have the stones to say this."

Engle has made similar statements at academic meetings, large and small, sometimes with Jaeggi and other targets of his attack sitting in the audience. After giving a speech at Rutgers University, for instance, he was asked by a student if he knew of anything that can increase working memory. "I have not seen a single good science demonstration that tells me that's the case," he answered. "Every study that I know of falls around the level that I call crap."

In one of his calmer moments, he told me: "Fluid intelligence is not culturally derived. It is almost certainly the biologically driven part of intelligence. We have a real good idea of the parts of the brain that are important for it. The prefrontal cortex is especially important for the control of attention. Do I think you can change fluid intelligence? No, I don't think you can. There have been hundreds of other attempts to increase intelligence over the years, with little or no—just no success."

I have always prided myself on being a skeptical bastard, which goes with the job description of being a science journalist. But I learned long ago not to be too easily swayed by either side in disputes of this nature, which sometimes resemble political conflicts in their seeming irresolvability. One of the first controversies I wrote about, back in the early 1990s, was over how best to treat prostate cancer. One group of researchers insisted that surgery and radiation are absolutely lifesaving. Those researchers, by the way, just happened to be surgeons and radiologists. Another group of researchers insisted that watchful waiting is the best option, because prostate cancer is usually slow growing, and treatments—whether surgery, radiation, or both—can have devastating side effects. Those guys happened to be epidemiologists. And guess what: twenty years later, they're still arguing with each other.

Science in general, and medicine in particular, has these kinds of disputes all the time. As in politics and sports, they're part of the game. Engle's insistence, therefore, that he knows the truth and that so many others are wrong cannot be taken as gospel, no matter that he was the one who put working memory on the map. After all, even Einstein got it wrong about certain aspects of quantum physics. So let's look at the evidence.

•—•

As stated in the introduction, as of this writing, I am aware of seventy-five randomized trials, published in peer-reviewed scientific journals, that have found a significant benefit to cognitive training of various sorts, and a grand total of four that have found no such benefit. A fifth study did not actually involve any working-memory training but merits attention because of the considerable press coverage it drew in Great Britain when it was published in June 2010. Neuroscientist Adrian Owen conducted the online experiment in coordination with the BBC television show *Bang Goes the Theory*. After inviting British viewers to participate, Owen recruited 11,430 of them to take a battery of online IQ tests before and after a six-week online program designed to replicate commercially available "brain building" software. Some of the participants were assigned to six "reasoning, planning and problem-solving tasks," while others trained on six "tests of memory, attention, visuospatial processing and mathematical calculations." The tasks were based on the types that have been sold since 2005 by Nintendo under the names Brain Age: Train Your Brain in Minutes a Day! and Dr. Kawashima's Brain Training. Long criticized as lacking evidence of effectiveness, the games are carefully marketed by Nintendo, despite their alluring titles, as purely for "entertainment" purposes. Owen concluded, in a paper published in *Nature*, that while participants improved over the course of six weeks on all the trained tasks, "no evidence was found for transfer effects to untrained tasks, even when those tasks were cognitively closely related."

Reached by telephone, Owen initially expressed the kind of skeptical certitude that Randy Engle would appreciate. "I'm absolutely sure brain training doesn't work," he began. "I see no evidence for it in the world. And millions of people are using these." But when I asked him about research into training with the N-back and other working-memory tasks, he said, "I think a lot of the work that Jaeggi has done is excellent. Her working-memory task is unlike anything commercially available. It's enormously difficult. I think Jaeggi and colleagues are chipping really diligently away at this question of whether you can train fluid intelligence. It's very likely you can, if you work out the components. Why wouldn't you be able to get better at fluid intelligence? I do think Jaeggi and her company have done some superb work. I do hope they keep at it."

One of the first studies to specifically test the effects of working-memory training and find no benefit whatsoever was titled "No Evidence of Intelligence Improvement after Working Memory Training: A Randomized, Placebo-Controlled Study." Six of its eight authors were current or former researchers at the Georgia Institute of Technology, better known as Georgia Tech, including the paper's senior author, professor of psychology and interim director of the university's Center for Advanced Brain Imaging, Randall W. Engle.

Engle and his acolytes began their paper by deconstructing Jaeggi and Buschkuehl's pioneering 2008 paper, going over many of the same points that he had presented at meetings. Its cardinal sin, in their judgment, was that it presented the combined results of four smaller studies, each of which varied in a number of ways, so that Jaeggi and Buschkuehl had really been mixing apples and oranges. Jaeggi and Buschkuehl's original paper had described those four smaller studies, however, without worrying their editors or independent peer reviewers; even the celebratory commentary that accompanied it acknowledged the paper's shortcomings. But then, faults of design and statistical strength are a nearly universal quality of pioneering studies like theirs, meant not so much to answer a question as to raise one. Dozens of studies, including some by Jaeggi and Buschkuehl themselves, have since confirmed and extended the findings of their 2008 study. But then, all of those studies were worthless, too, in Engle's not-so-humble opinion.

Engle then described his own team's experiment: 130 participants were recruited, but only 75 of them completed the study and were included in the final analysis, all of them between the ages of eighteen and thirty, most of them college students, plus a small number of nonstudents. Participants were randomly assigned to one of three groups: one group to do the adaptive dual N-back task that Jaeggi and Buschkuehl designed, in which it would get harder as a person's skill improved, for twenty sessions; an "active control" group, in which participants trained on an adaptive visual search game that had no working-memory component, also for twenty sessions; and a no-contact control group that received no practice and had to show up only for testing. Three times during the study—before the training sessions began, midway through, and at the conclusion—all 75 participants underwent exhaustive testing of both their fluid and their crystallized intelligence, involving fourteen tests in all, including two versions of the Raven's, tests of the ability to understand inferences and analogies, three tests of multitasking, two of working memory, two of perceptual speed, one of vocabulary, and one of general knowledge. On average, the pretesting took two hours and twenty minutes, while the mid- and posttesting took an hour and forty minutes each. Participants were paid $40 each time they finished one of the testing sessions, plus a $12 bonus if they completed all three.

The researchers found no benefit to training whatsoever on any measure. Oddly, however, none of the groups showed evidence of the so-called testing effect, the general tendency for people in studies like this to get better on any given test as they repeat it. The testing effect explains why Kaplan prep courses devote much of their time to having students take and retake simulations of the exam they're prepping for: folks generally score higher on the second and third go-arounds with these tests for no other reason than that they get more familiar with them. That's the whole point of having placebo control groups in training studies, so that researchers can separate out the testing effect, which enables almost everyone to do a little better on a follow-up test, from the gains specifically attributable to training. The fact that Engle's participants scored about the same on their pretest, midtest, and posttest is just odd, and it raises questions: Were participants getting exhausted by having to take such a lengthy battery of tests—fourteen of them—on three separate occasions? Was there too little room for improvement with so many tests given in a couple of hours?

But let's give Engle the benefit of the doubt that he does not give to Jaeggi, Buschkuehl, and others. Really, every study ever designed, including those by physicists but especially by psychologists, is subject to carping, critiquing, and second-guessing. So let's accept at face value that Engle's study was published in a peer-reviewed journal and that it found no benefit from N-back training under the conditions the researchers designed.

Another paper to find no improvement in intelligence from working-memory training in healthy young adults was titled, strangely enough, "Working Memory Training Does Not Improve Intelligence in Healthy Young Adults." Weng-Tink Chooi, of the University of Science in Malaysia, and Lee A. Thompson, a psychologist at Case Western Reserve University in Cleveland, started out with 130 undergraduate students who showed up for pretesting before the training began. They each took a vocabulary test, a perceptual speed test, a few tests of visuospatial reasoning, and the Raven's progressive matrices. The students were told they would earn four to six academic credits for their first six hours of participation and $7.50 per hour for each subsequent hour in the study.

They were then randomly assigned to one of six groups. The first group did the dual N-back for eight days, each day spending about a half hour. Another group did the same, but for twenty days. Both of these groups followed the protocol that Jaeggi and Buschkuehl had designed, in which the degree of difficulty adapts to the person's ability, rising to 3-back and 4-back and beyond as the person's accuracy grows. The third and fourth groups did the dual N-back for eight or twenty days, but remained stuck at 1-back the entire time. As boring as 1-back must have been, these participants were considered "active" controls because they were at least actively doing something. The fifth and sixth groups, considered the "passive" controls, showed up for eight or twenty days but were given no tasks whatsoever; they could sit and do homework or nothing at all. At the conclusion of the study, 93 of the original 130 students showed up for posttesting and were included in Chooi and Thompson's analysis.

The results were peculiar: not only did the students in both the eight- and the twenty-day groups doing the progressive dual N-back not improve on most of the follow-up tests, they actually got worse on most of them. The same drop was seen on most of the follow-up tests taken by all the control groups. This reverse-testing effect suggests that something went a little funky (or, as programmers like to say, kludgy) in the study's implementation. Another peculiarity of Chooi and Thompson's study is that the students' assignment to active or placebo treatment groups was not strictly random: if those initially assigned to do N-back didn't feel like it, they could opt out and switch into the do-nothing group. As the authors acknowledged in the paper, "We speculated that only the less motivated participants would have opted-out, and they may be the very individuals who could have benefited from the training and led to significant transfer effects." (Jaeggi and Buschkuehl's 2008 study, and many others, had found that those who begin at the lower end of abilities tend to gain more, proportionally, than those at the higher end, if for no other reason than that they have so far to go.) As a result of this opting out, some of the training groups had as few as ten participants, which limited the study's statistical power.

But again, as with Engle's study, we must concede that Chooi and Thompson's study was published in a peer-reviewed journal and found no benefit to N-back training.

The third study to find no benefit was published on May 22, 2013, in the open-access journal *PLOS ONE.* Funded by the Defense Advanced Research Projects Agency (DARPA) and led by John Gabrieli, a professor in the department of brain and cognitive sciences at the Harvard-MIT Division of Health Sciences and Technology, it involved fifty-eight college-age adults recruited from around the MIT campus. With an average IQ of 117 to 120, the volunteers were paid $20 for each of twenty training sessions, and more for completing all sessions and undergoing before-and-after testing. At the study's conclusion, the twenty volunteers who had been assigned to adaptive dual N-back training showed negligible improvements on most of the untrained tasks, and nothing statistically better than the volunteers assigned to active or passive placebo—not on other working-memory tasks and not on measures of fluid intelligence.

Perhaps the high IQs of the MIT-area volunteers was the problem; as just noted, some studies have found the greatest benefit in people most in need of it. Other than that, I have to admit that this one, of the four studies showing zero benefit, is the one that gives me the willies.

The fourth study showing no benefit of training is by far the least impressive, published in an obscure journal that is not indexed on PubMed, *Computers in Human Behavior.* It involved thirty-nine people randomly assigned to one of four groups: one playing Dr. Kawashima's Brain Training, another playing a strategy game, a third playing a version of the dual N-back created by the investigators, and the fourth serving as a passive control. After an average of seventeen sessions lasting twenty minutes each (which the participants were instructed to carry out at their homes), the dual N-back group increased its score on the Raven's advanced progressive matrices more than any other group, but not enough to be statistically significant. Really, this one should be marked "Return to Sender."

Still, taken as a group, if these four were the only studies out there, they would be pretty damning. But then there are seventy-five other studies showing benefits, including twenty-two showing increased fluid intelligence. The question, then, is how to view the negative results, given that it's a messy world out there, especially when attempting to measure anything as slippery as human intelligence. The answer is that, like a jury, we must weigh the totality of the evidence.

Engle has been good enough to try to weigh it for us. In the third issue of the brand-new *Journal of Applied Research in Memory and Cognition*, for which he just happened to serve on the editorial board, Engle and two of his Georgia Techies published a paper entitled "Cogmed Working Memory Training: Does the Evidence Support the Claims?" They reviewed twenty-one previously published studies of Cogmed, beginning with the positive results reported by Klingberg in his seminal 2002 paper, the one that caught the notice of Jaeggi and Buschkuehl. "Although these results were encouraging," Engle and colleagues wrote, "a series of failed replications followed."

Gee, that sounds disappointing. But wait a minute. Here are the titles of the next three studies Engle cites as examples of "failed replications":

"Working Memory Deficits Can Be Overcome: Impacts of Training and Medication on Working Memory in Children with ADHD."

"Adaptive Training Leads to Sustained Enhancement of Poor Working Memory in Children."

"Effects of Working Memory Training on Reading in Children with Special Needs."

That last study, led by psychologist Erika Dahlin, involved fifty-seven Swedish primary-school children with special needs. It concluded, after providing forty-two of them with Cogmed training and using the rest as comparisons: "The results show that working memory can be seen as a crucial factor in the . . . development of literacy among children with special needs, and that interventions to improve working memory may help children [become] more proficient in reading comprehension."

Almost every other published study Engle cites as evidence that Cogmed doesn't work found evidence that it actually did, in one respect or another. In fact, I count many of those studies, including the three listed above, as evidence that working-memory training *does* work. What Engle does—and it's interesting that he used the term "cherry-picking" in regard to Jaeggi and Buschkuehl—is to mention only the negative test results in each study, ignoring the many positive results. A few other of these "failed" studies:

"Working Memory Training for Children with Cochlear Implants: A Pilot Study."

"Computerized Working Memory Training Improves Function in Adolescents Born at Extremely Low Birth Weight."

"Changes in Cortical Dopamine D1 Receptor Binding Associated with Cognitive Training" (published, by the way, in *Science*, one of the top scientific journals in the world).

"Increased Prefrontal and Parietal Activity after Training of

Working Memory" (published in another highly regarded journal, *Nature Neuroscience*).

Somehow, after reviewing these studies, Engle had the stones, to borrow his phrase, to conclude that "the claims made by Cogmed are largely unsubstantiated" and that "for people seeking increased intelligence, improved focus and attention control, or relief from ADHD, current research suggests that this training program does not provide the desired result."

So bizarre was this conclusion, given that Cogmed is without question the most-studied commercial service for training working memory, that the rest of the issue in which Engle's paper appeared was taken up by responses from other researchers. None of them went so far as to call Engle's review a hatchet job, but some came close. Susan E. Gathercole, a psychologist at the University of York, England, coauthored two of the studies that Engle critiqued. While conceding that both of those papers had been relatively small, she stated that "the consistency of the enhancement in working memory performance that we had seen" gave her group confidence to launch a much larger, longer, more expensive, and more elaborately designed study. The results, she said, are "upholding our previous findings." Without the earlier studies, she noted, the larger study "would not have been justified." She concluded with a slap at Engle's harsh judgments: "In evaluating research on cognitive interventions, it is important to weigh up all the available data. Cumulative evidence, gauged appropriately, can have immense value. A skeptical stance is vital in science, but it is important also to avoid throwing out the baby with the bathwater."

A group from the University of Notre Dame published two papers in the same issue. The first, a randomized study, concluded that Cogmed training does not appear to improve people's ability to draw from their long-term memory while trying to solve working-memory problems. The group's second paper, however, emphasized that "the full potential of Cogmed working memory training remains unknown at this time" and suggested that it could well be modified to improve more aspects of working memory.

Jaeggi, Buschkuehl, and some of their former colleagues from the University of Michigan also published responses in the issue, conceding that the case for Cogmed may not be "airtight" but calling Engle's criticisms "overly pessimistic." Klingberg, too, contributed a paper, the final paragraph of which gets to the heart of the matter: "Working memory training is still a young field of research. As with all science, no single experiment explains everything, and results are never perfectly consistent. . . . Many questions remain. But there is no going back to the notion that working memory capacity is fixed."

The only paper in the issue that shared Engle's nihilistic views on Cogmed was coauthored by a colleague of Gathercole's at the University of York, Charles Hulme, professor of psychology. He and Monica Melby-Lervåg, a professor in the department of special needs at the University of Oslo, Norway, presented a statistical meta-analysis of the same papers cited by Engle, concluding that overall, "evidence for the effectiveness of the Cogmed program is extremely weak." By this, what Hulme and Melby-Lervåg really meant was that, statistically speaking, most of them were either not big enough or not strong enough to convince them. So even though every paper they reviewed had presented *modest* statistical evidence that Cogmed's working-memory training does provide benefits, in Hulme and Melby-Lervåg's upside-down version of statistical reality, all those studies actually prove the opposite.

For all the gusto with which Engle's review, and Hulme's meta-analysis, sought to poleax Cogmed, they failed to convince many of their peers in academia, or any that I could find. Quite the opposite: at least fifty-seven other studies of Cogmed are under way, including

Gathercole's. As noted in chapter 3, Kristina Hardy's study of childhood cancer survivors and Julie Schweitzer's of children with ADHD both reported benefits of Cogmed last year, within months of Engle's and Hulme's attempts at mass deconstruction. And in June 2013, the *American Journal on Intellectual and Developmental Disabilities*, the leading journal in its field, published the results of Stephanie Bennett's study of twenty-one children with Down syndrome. The children, aged seven to twelve years old, were randomly assigned either to serve as controls or to spend ten to twelve weeks training on a preschool version of Cogmed for twenty-five minutes per day at their local school. "Following training," the study concluded, "performance on trained and untrained visuospatial short-term memory tasks was significantly enhanced for children in the intervention group. This improvement was sustained four months later. These results suggest that computerized visuospatial memory training in a school setting is both feasible and effective for children with Down syndrome." Brian Skotko, codirector of the Down Syndrome Program at Massachusetts General Hospital, told me, "If Cogmed was a drug, everyone would call this study groundbreaking."

"I was a hillbilly kid from West Virginia."

Randy Engle turns out to know a thing or two about overcoming the limitations of birth and background. Sitting in the cafeteria at Rutgers University, he looks younger than his sixty-six years, with a burly build, ruddy cheeks, and a dwindling tribe of hairs clinging to their position at the center of his high forehead.

"I grew up in the Kanawha Valley in West Virginia. Until fourth grade, I lived in a house with no inside plumbing. I was one of four kids. My mom used to put us in a number-nine washtub, the biggest tub you could get. She would heat water on a woodstove. She cooked over a woodstove. And I went to a one-room school from first to eighth grades.

"But I'm proud of my background. I consider myself fairly privileged, because my mom, like the parents of most of the ones who got out of a condition like that, my mom believed in me, she didn't push me, but she enabled me. My dad, on the other hand, just didn't have a clue. He had to drop out of school during the Depression to support his family. He was fifteen. He ended up becoming a factory worker. He was incredibly smart, but my mother had the vision that we could be something else. I was the first high school graduate on his side. First one. I was the first one to go to college on either side. Most of my siblings went to college, too.

"But I had no script. I lived six miles away from an 'HBCU'—that's an historically black college or university. West Virginia State was the HBCU. I was a minority there. It was 75 percent black. I got a great education. There was a large faculty of teachers who couldn't get jobs elsewhere because of the color of their skin. I had a psychology professor who got his PhD from Northwestern in 1929. Worked as a butler through graduate school. I had a math professor who was a Harvard PhD, who would have been hired anywhere else in the modern era. I took French from a woman who got her PhD at the Sorbonne in the 1930s. Just a tremendous undergraduate education.

"West Virginia, you see, was functionally a colony not so long ago. The people who owned the coal mines were up north. Nobody bought their own house; the coal mines owned the land. And they paid the people in scrip, not American coinage. The only place scrip could be used was in the company store. So it was really important that the employees not be very educated. Let that sink in. It was really important that they had a cheap, controllable workforce. These owners and managers put people under working conditions that were terrifically hazardous to their health. In order to keep people working fifty weeks a year, eight to ten hours a day, as long as they lived, you had to keep them ignorant. The other thing those owners used was race. If I can convince you that your condition, bad as it is, will be worse if you let 'these people' in, these immigrants, these blacks, then you'll defend the system. So racism, it was really important to keep that alive.

"I gave every nickel I legally could to Obama. My son-in-law is a Republican. I told him, 'I am spending your inheritance to make sure your candidate doesn't win.' Because I have grandchildren, and I don't want them to grow up in the kind of plutocracy like the Bushes or the Romneys or the Kochs or the Adelsons would like to have."

Back on the subject of intelligence research, Engle asked me, "Do you know who Rushton was? He was the worst." (J. Philippe Rushton, a psychologist at the University of Western Ontario, became famous in the 1980s and 1990s for espousing racist views on the genetics of intelligence and crime.) "But there's an absolute bias in American psychology toward a strict nurturist position and against any effect of inborn nature," Engle continued. "Any behavioral geneticist worth ten cents knows that fluid intelligence is about 50 percent inherited. I would never say that nurture and environment are unimportant. In my view they're huge in overcoming limits on native intelligence. I don't think I'm very smart but I work my ass off. But can you make your spleen bigger? I don't think so. I'm not saying there's no way intelligence can increase. But to think you can do that in ten hours, give me a break. I say to any serious psychological scientist, that's an absurd thing to believe."

As Engle defended the importance of biologically based differences between people in their level of fluid intelligence, I wondered what he makes of the views of J. Anders Ericsson, the psychologist known for his view that experts and even so-called geniuses are no different from anybody else except in the amount of time they have devoted to their chosen specialty and the amount of specialized knowledge they have thereby gained. Ericsson has famously argued that whether in chess or playing the violin, the top achievers are the ones who have devoted an average of ten thousand hours to gain the most knowledge and master their profession. In *Outliers*, Malcolm Gladwell called this the "ten-thousand-hour rule."

"Anders and I have this running argument," Engle said. "For Anders knowledge is everything and traits are nothing. It's John Locke cubed. But Zach Hambrick has some really interesting studies where he shows that working-memory capacity accounts for a lot of the variance in skills at Texas hold 'em, piano, you name it. Zach's work is amazing. So I'd rather be knowledgeable than ignorant, but there's a role for trait ability above and beyond knowledge."

Ericsson dismisses the notion that fluid intelligence can be trained not because he shares Engle's view that it's both important and unchangeable, but because he believes it's *unimportant* and unchangeable.

"Given our work, we don't really see the importance of generalized capacities," Ericsson told me in a telephone call from his office at Florida State University in Tallahassee, where he is the Conradi Eminent Scholar and professor of psychology. "My main problem with Susanne Jaeggi's work is that she seems not to be content with accepting that you can improve performance on the task you're training on. She wants to show this improves your intelligence generally. What really annoys me is that by proposing to train intelligence, I think they're denigrating what we've learned about the prerequisites for achieving high performance in any domain. If you could find even a single case where, with fifteen hours of practice, you could produce a world-class expert, that would violate this ten-thousand-hour rule and everything we've found, that even the most gifted has to spend thousands of hours of deliberate practice to get there."

Of course, neither Jaeggi nor anyone else involved in cognitive training has ever suggested that a person could become a world-class expert in fifteen hours of working-memory training. What they do assert is that training permits people to learn faster, so that they might need less time to achieve mastery on any given subject. But if it's really true, as Ericsson asserts, that practice, practice, practice is the only thing dividing the expert and nonexpert, no matter a person's cognitive capacities, then cognitive training doesn't matter, intelligence doesn't matter, nothing matters except those ten thousand hours of practice. Of course practicing one piece of music will make you better at learning other pieces of music, but beyond that, in Ericsson's view, there is no transfer.

Hambrick, who obtained both his master's and his PhD in psychology under Engle, now serves as associate professor of psychology at Michigan State University. Hambrick has debated me on Wisconsin Public Radio over the trainability of intelligence, and he coauthored both of Engle's papers bashing the field of cognitive training, but he has some fascinating critiques of Ericsson's claims.

"Ericsson has this strong argument that it's all about practice," Hambrick told me. "Aaaaaand so here's the thing about that. When you look at expert musicians, expert chess players, expert Scrabble players, expert pianists, there is a tremendous amount of variability among those experts in their estimated amount of practice. Ericsson has ignored the variability in his own data for reasons I don't quite understand. Among chess masters, the amount of time they spend practicing before reaching that level ranges from about two thousand hours to about twenty thousand hours. We see this in Ericsson's own data. He had a study of Scrabble players. Of course it was the case that the experts had practiced longer, on average, than the novices.

But there's tremendous variability among those experts. It's not the case that they all had ten thousand hours of practice. The individual differences in practice among experts is enormous. It varies by orders of magnitude."

I mentioned Steve Winwood, the guitar prodigy who joined the Spencer Davis Group when he was fourteen, back in the early 1960s, was recording with Eric Clapton when he was sixteen, and had formed both Traffic and Blind Faith before turning twenty-one. The example of Winwood had always been a burr in my fur, because I'd practiced guitar with religious zealotry in my teens and never achieved more than a crazy punk band, the Mutations, in college. To my surprise, Hambrick knew all about Winwood.

"Not only was he an expert guitarist at a very young age," he said, "he was an expert on a lot of different instruments. He played all the instruments on his solo albums."

Now Hambrick was getting jazzed.

"Let me read you a paper I'm writing," he said. "In one sample of chess players—these are master-rated players, with a rating of at least 2200—their hours of deliberate practice before getting that rating ranged from less than a thousand hours up to twenty-four thousand. What's even more critical than that is, when we compare a master group to a novice group or intermediate group, there are some master-level players who have less practice than the average intermediate players. There are some intermediate players who have practiced longer than the average master player. The point is: there's a huge amount of variance." That paper he was writing, eventually published online in May 2013 in the journal *Intelligence*, concluded that deliberate practice accounted for only about one-third of people's achievements in both chess and music.

"Gladwell in *Outliers* says, 'research suggests that once a musician has enough ability to get into a top music school, the thing that distinguishes one performer from another is how hard he or she works.' That's just not correct," Hambrick said. "There's a tremendous amount of variability in how hard people have to practice. The ten-thousand-hour rule is—I think it's a myth. 'Researchers have settled on what they believe is the magic number for true expertise: ten thousand hours.' That's an average number, but there's enormous variability around those averages. This is a really important issue. How do people become great at what they do? Of course practice is important, but it's not everything."

As Engle told me, Hambrick has even tested the effects of practice on success in poker. In 2012, he coauthored a study of 155 Texas hold 'em players of widely varying experience. They found that experience counted for a lot—57 percent of the variance, to be exact, in their ability to evaluate a winning hand, and 38 percent of their ability to recall which cards had been shown so far. However, they wrote, "working memory capacity added significantly to the prediction, and there was no evidence for interactions between poker knowledge and working memory capacity. That is, working memory capacity was as important as a predictor of performance at high levels of poker knowledge as at low levels, suggesting that domain knowledge may not always enable circumvention of working memory capacity in domain-relevant tasks." Working memory accounted for 19 percent of a person's ability to evaluate a winning hand, they found, and about 32 percent of his or her ability to remember the cards already shown in a game.

•—•

Much as I enjoyed talking about rock guitarists and poker players with Hambrick, and much as I like and respect him and Engle both, I still did not understand why they have taken such a stand against the large and growing body of evidence showing that working memory and fluid intelligence can be increased through training. But I kept trying.

"I'm simply not convinced that working-memory training has generalizable benefits," Hambrick told me.

But if a person trains and improves on working-memory tasks, I asked, doesn't that mean their working memory is better? And isn't that valuable?

"It could be," he conceded. "Let's say you practice a particular working-memory task over and over, and you get really good at it. Then the question is, does that have any benefits beyond that particular task? That's where the real controversy is. It does stand to reason that if working memory is an important component of fluid intelligence, and if it's predictive of fluid intelligence, then if you get better on a test of working memory, you've improved your fluid intelligence. Or maybe this whole notion that working memory causes fluid intelligence is wrong. Which is a bit unsettling, because we've spent a lot of time trying to make the case that working memory and fluid intelligence are correlated, and have argued that the causal arrow goes from working memory to fluid intelligence. But, you know what, we could be wrong."

Then the direction of his thinking changed. "Maybe people can acquire working-memory skills that are generalizable beyond the training task," he said. "And I think that's an exciting possibility. So maybe people would be better at doing mental arithmetic, as a possibility. But that's very different from saying that working-memory training enhances fluid intelligence."

This kind of wishy-washy admission that working-memory training might improve some highly useful aspects of cognitive functioning, but not fluid intelligence, makes little sense given Hambrick's and Engle's career-defining view that the two are so closely related as to be difficult to tease apart. Yet at various times in my talks with Engle, even he went wishy-washy on me.

"I do think we can all learn to attend better," Engle once told me. "I spent a whole summer one year reading books on mindfulness. The Zen Buddhists have a better idea of mindfulness than I do, and how to train it. Yes, I do think that that aspect of working memory, and to the extent it plays a role in fluid intelligence, that it can be improved. I think there are limits to how much it can be improved. That capability is driven by brain structures, by genetics. But it can be improved quite a bit. My three-year-old grandson, when he comes to visit me, we go into my front yard. We started this game a couple of years ago. We sit on a rock wall I have in my front yard. I'll say, 'I hear a bird. Do you hear that bird?' And he listens. Then I'll say, 'I hear an airplane. Do you hear an airplane?' And he's gotten better at it. He's developing skills for separating the signal from the noise."

Of Julie Vizcaino's improved grades and apparently improved intelligence after she devoted so much time to mastering chess, he said, "That probably has generalized to a great deal of other aspects of her life. I believe that. I do think that there are abiding individual differences in brain structure. But—and it's a big but—I think we can learn to circumvent those, to minimize those limitations. I think learning to pay better attention, to block out distractions, to not let your attention be distracted by irrelevant events, can help you overcome your limitations. It makes those limitations less important. To be honest with you, I'm a living example of that. I've never thought of myself as very smart. But I've had the advantage of just being stubborn as hell and persisting. So I've had some level of success in my life because I didn't let limitations get in the way. It's the equivalent of giving a short guy a ladder. And there are lots of ways we as a society can do that. I am interested in those limitations, but I am also interested in how we can get around those limitations."

During another conversation, instead of the metaphor of "giving a short guy a ladder," he related the effects of training to a scene from director James Cameron's *Aliens*, in which Ripley, the character played by Sigourney Weaver, climbs into a powered exoskeleton designed for loading cargo in order to battle the giant Alien queen.

"So we can put Sigourney Weaver into this robot suit and all of a sudden she can lift one hundred times more weight than she could before," he said. "What we've done is helped her circumvent her own limitations. You and I do this all the time. That's not doing away with our limitations. It's just finding a way around them."

But when we raise a ladder inside our brain, who can tell the difference anymore between what we were and what we have become?

CHAPTER 9

Flowers for Ts65Dn

If the promise of increasing fluid intelligence is really a fiction, as Engle believes, then all the published studies might as well be placed on a bookshelf next to the first, great work of fiction on the subject: *Flowers for Algernon*, the 1966 novel by Daniel Keyes. Originally published as a short story in the April 4, 1959, issue of *The Magazine of Fantasy & Science Fiction,* and later made into the movie *Charly*, the novel narrates the story of a mentally disabled man whose intelligence is temporarily increased to genius level following an experimental brain surgery. Algernon is the laboratory mouse who first undergoes the surgery and then shows a temporary, dramatic improvement in his maze-learning abilities.

But what if Keyes's science fiction can be turned into scientific fact? The closest any scientist has yet come to doing so is a Brazilian physician and neuroscientist named Alberto Costa. Early in the evening of June 25, 1995, hours after the birth of his first and only child, the course of his life and work took an abrupt turn. Still recovering from a traumatic delivery that required an emergency cesarean section, Costa's wife, Daisy, lay in bed, groggy from sedation. Into their dimly lit room at Methodist Hospital in Houston walked the clinical geneticist. He took Costa aside to deliver some unfortunate news. The baby girl, he said, appeared to have Down syndrome, the most common genetic cause of cognitive disabilities, or what used to be called "mental retardation."

Costa had only a minimal knowledge of Down syndrome; it wasn't his specialty. Yet there in the hospital room, he debated the still-tentative diagnosis with the geneticist. The baby's heart did not have any of the defects often associated with Down syndrome, he argued, and her head circumference was normal. She just didn't look like a typical Down syndrome baby.

"But we have the blood sample," Costa recalls the geneticist telling him. That, he knew, meant a genetic test had already been done. Diagnosis is made when a child is born with three copies of all or most of the genes on the 21st chromosome, instead of the usual two copies.

Costa had dreamed that a child of his might grow up to become a mathematician. He had even prevailed upon Daisy to name their daughter Tyche, after the Greek goddess of chance and in honor of the Renaissance astronomer Tycho Brahe. Now he asked the geneticist what the chances were that Tyche (pronounced "Tishy") really had Down syndrome.

"In my experience," he said, "close to a hundred."

As distressing as the news was, it did not come entirely out of the blue. Daisy's first pregnancy had ended in a miscarriage, which, they knew, can occur because of a genetic disorder in the fetus. When Daisy got pregnant a second time, Costa insisted they get a chorionic villus sampling, the definitive prenatal genetic test. But the procedure caused a miscarriage. (The test showed that the fetus was genetically normal.) Guilt-ridden, Costa vowed that if there was a third pregnancy—this one—they would conduct no prenatal tests.

Now, with Tyche bundled peacefully in a bassinet at the foot of Daisy's bed, and Daisy asleep, Costa sat up through most of the night crying. He had actually gone into research to get away from the practice of medicine precisely because scenes like this—parents devastated by a diagnosis—had proved too painful for him. But by morning, he found himself doing what any father of a newborn might: hovering by the crib, holding his daughter's hand, and marveling at her beauty.

"From that day, we bonded immediately," he told me during one of our many talks. "All I could think was, she's my baby, she's a lovely girl, and what can I do to help her? Obviously I was a physician and a neuroscientist who studies the brain. Here was this new life in front of me and holding my finger and looking straight in my eyes. How could I not think in terms of helping that kid?"

With no experience in the study of Down syndrome, Costa took a short walk the next day to the library affiliated with Baylor College of Medicine, where he worked as a research associate in the division of neuroscience. Reading the latest studies, he learned that the prognosis was not nearly as dire as it had once been considered. Life expectancies had grown, educational reforms had produced marked gains in functioning, and—of particular interest to Costa—a mouse model of the disorder had recently been developed, opening the door to experimentation. By day's end he had made a decision: he would devote himself to the study of Down syndrome.

The results would electrify the field. In 2006, using mice with the equivalent of Down syndrome, Costa published one of the first studies ever to show that a drug could normalize the growth and survival of new brain cells in the hippocampus, a region deep within the brain that is essential for memory and spatial navigation. In people with Down syndrome, the slower pace of neuron growth in the hippocampus is hypothesized to play a key role in their cognitive deficits.

Follow-up studies by other researchers reached conflicting results as to whether the drug Costa had tested, the antidepressant Prozac, could produce practical gains on learning tests to match its ability to boost brain-cell growth. But, undeterred, Costa moved on to another treatment strategy. In 2007 he published a study that showed that giving mice with Down syndrome the Alzheimer's drug memantine (marketed under the brand name Namenda) could improve their cognitive performance.

The following year, Costa took the next great step, launching the first randomized clinical study ever to take a drug that worked in mice with Down and apply it to humans with the disease. Neuroscientists in the field told me that however it turned out, the study marked a milestone.

"This was a disorder for which it was believed there was no hope, no treatment, and people thought, why waste your time?" said Craig C. Garner, professor of psychiatry and behavioral sciences and codirector of the Center for Research and Treatment of Down Syndrome at Stanford University. "The last ten years have seen a revolution in neuroscience, so that we now realize that the brain is amazingly plastic, very flexible, and systems can be repaired."

But other scientists were busy pursuing an entirely different kind of breakthrough: rather than treat Down syndrome, they wanted to prevent it. Noninvasive prenatal blood tests were being developed to allow for routine testing in the first trimester of a pregnancy, raising the specter that many more parents would terminate an affected pregnancy.

"It's like we're in a race against the people who are promoting those early screening methods," Costa told me. "These tests are going to be quite accessible. At that point, one would expect a precipitous drop in the rate of birth of children with Down syndrome. If we're not quick enough to offer alternatives, this field might collapse."

•—•

London physician John Langdon H. Down was the first to describe the disorder that would eventually bear his name, in a remarkable paper published in 1866 under the title "Observations on an Ethnic Classification of Idiots." At the time, "idiot" and "imbecile" were accepted medical terms to describe the intellectually disabled living among the inhabitants of Dickens's England. Dismayed by his profession's failure to distinguish among the various types and causes of such disabilities, Down proposed that they could be classified based on what he considered to be a resemblance to various ethnicities: Caucasian, Ethiopian, Malay, and—infamously—Mongolian. Amid the welter of well-intentioned hooey, Down accurately noted that people he called "Mongol" (a term that eventually morphed into the epithet "Mongoloid") had the disorder from birth, not due to any injury or illness; that they had "long, thick" tongues, difficulty speaking, a "flat and broad" face, a heavy-lidded appearance to the eyes, abnormal physical coordination, and a cheerful disposition. The feature he mentioned most often, however, was that they benefited greatly from training and instruction.

That last point appears to have been largely forgotten by 1959, when French geneticist and pediatrician Jérôme Lejeune proved that the disorder is caused by a third copy of the 21st chromosome, rather than the usual two copies (one inherited from each parent). So recently was Lejeune's discovery made that in March 2011, Costa actually met his widow. The scene of their meeting was a Paris conference, named in honor of Lejeune, where neuroscientists from around the world discussed progress into treatments for Down and related diseases. Such a conference would have been inconceivable when Costa entered the field fifteen years earlier.

"If you think about most genetic diseases, they're caused by one gene, and in fact one mutation at one amino acid," said Roger Reeves, a leading researcher in the field and professor at the Institute for Genetic Medicine at Johns Hopkins University School of Medicine. "With Down syndrome, you have an extra copy of all five hundred or so genes on chromosome 21." For decades, Lejeune's discovery served to scare off scientists from any serious effort to find a medical treatment for what they were soon calling "trisomy 21." It just seemed impossibly complex.

"The turning point really came when Muriel Davisson made her mouse," Reeves told me.

Davisson, now retired from a long career as a geneticist at the Jackson Laboratory in Bar Harbor, Maine, spent years breeding a mouse—an Algernon, if you will—that would have many of the traits associated with Down syndrome. The task was complicated enormously by the fact that the genes that occupy the 21st chromosome in humans are spread out, higgledy-piggledy, across the 10th, 16th, and 17th chromosomes in the mouse. So what did she do? Three words: Male. Gonads. Irradiation.

"It's known that if you irradiate the gonads of male mice, the chromosomes will break and sometimes, randomly, fuse back together with the wrong chromosome," Davisson told me. If extra copies of the Down genes would happen by chance to bunch up onto a single chromosome following irradiation, she figured, she just might hit the jackpot. With funding from the National Institute of Child Health and Human Development, an arm of the National Institutes of Health, she began the process of irradiating the gonads of male mice in 1985.

Five years and 250 mice later, she concluded that the sixty-fifth attempt was her keeper, with the best combination of Down-like qualities, even including, incredibly, some of the distinctive facial characteristics associated with the disease, and the same slightly uncoordinated gait. The mouse has since been known as Ts65Dn, for trisomic, 65th attempt, Davisson.

Five years after publishing news of her mouse, Davisson received an e-mail from a young neuroscientist named Alberto Costa. Her work, he told her, had opened the door for him to conduct meaningful new drug research.

"It was an epiphany, that, oh, this is a field where I can apply a lot that I've learned," Costa said. "Science is usually unforgiving with people who try to change career paths, but it was a risk I was willing to take." Having earned his PhD studying the electrical and chemical basis of communication between brain cells, he told me, "I figured, okay, if there is something that can be done in this field, it's going to be done at that level of neuronal electrophysiology." After months of reading the latest studies, Costa decided he needed to work with Davisson's mice in Maine.

"He twisted my arm till I took him into my lab," Davisson said with a laugh. "I didn't have funding. He wrote a grant to get the funding. He is very enthusiastic." But, she also found out, "He's a perfectionist, and not very tolerant of people who aren't perfectionists. He doesn't do experiments without being sure he's doing them right. When he makes a finding, you know that it's real."

Using Davisson's mice, Costa's 2006 study with Prozac helped set off a gold rush by other scientists panning for a drug that would produce not merely cellular changes in the brain but behavioral changes indicative of increased intelligence. First out of the gate was Craig Garner at Stanford, who in April 2007 reported behavioral improvements in Ts65Dn mice, but only after weeks of treatment with the experimental drug he tested. (In 2013, a company he cofounded to pursue that strategy received $17 million from a venture capital firm.) Four months later, Costa published his memantine study, which found that a single injection of the drug produced behavioral benefits within minutes, enabling Down-equivalent mice to master a water maze as quickly as standard mice.

Memantine works, Costa hypothesizes, not by boosting the growth of brain cells but by normalizing how existing cells use the neurotransmitter NMDA. Because people with Down syndrome have three copies of all or most of the genes on chromosome 21 instead of just two, they have about fifty percent more of any proteins encoded for by that chromosome. One result is that the NMDA receptors of Ts65Dn mice are "hyperactive"—they overreact to stimuli. By responding too easily, they learn too little; the signal is lost amid the noise. But giving memantine to quiet the noisy NMDA receptors, Costa has found, makes the brain cells react normally.

Other defects have also been observed in the Ts65Dn brain, and at least three drug studies by other researchers have since shown benefits in the mouse. In November 2009, William C. Mobley, chair of neurosciences at the University of California at San Diego and one of the most active and visible researchers in the field, coauthored a study showing that a combination of drugs designed to raise norepinephrine levels in the brain normalized the mice's learning abilities. More recently, in June 2010, Nobel laureate Paul Greengard of Rockefeller University entered the fray, showing that memory and learning could be normalized in Ts65Dn mice by lowering levels of beta-amyloid, the protein goop that has long been known to clog the brains of people with Alzheimer's disease.

"There's been a sea change in our ability to understand and treat Down syndrome," said Mobley. "There's just been an explosion of information. As recently as the year 2000, no drug company would possibly have thought about developing therapies for Down syndrome. I am now in contact with no less than four companies that are pursuing treatments."

Furthest along in translating mouse research into humans, Costa began his clinical trial in July 2008. His goal was to recruit forty young adults with Down syndrome. Half would receive memantine, the other half a placebo. After sixteen weeks of taking their assigned pills, they would all be retested to see whether those who had taken memantine had become, in a word, smarter.

This book never would have been written—I never would have been drawn so deeply into the subject of cognitive enhancement—if Costa had just shut up and stopped talking the first time I got him on the telephone. That was in December 2009, when Fay Ellis, my editor at *Neurology Today*, the official newspaper of the American Academy of Neurology, asked me to write an article about Mobley's mouse study. Mobley happened to mention that some guy in Colorado was actually running a clinical trial of a drug in young adults with Down syndrome. I called Costa for the usual ten-minute telephone interview, hoping to get a few quotes about Mobley's study and a little about his own, but he kept talking and talking for over an hour. Usually this is a journalist's worst nightmare, but the more Costa talked, the more interested I became.

"It's a field that's actually in the middle of what could be a real revolution," Costa began, "but nobody wants to hear about it. There's such apathy from the National Institutes of Health ever since chromosome 21 was sequenced. What I hear is: 'Why bother with Down syndrome? It's too complex, a whole extra chromosome, and we already have a response to it.' With the new prenatal tests, that's what the geneticists are expecting. 'It's going to disappear, so why fund research into treatments?' We have over 300,000 people with Down syndrome living in the United States right now. They may have the same fate as the dinosaurs.

"But Down syndrome isn't cool. Everything today is autism spectrum disorder. Even fragile X syndrome gets more funding and better press. The only time you see an article in the newspaper about Down syndrome is when they're talking about the prenatal tests.

"Even many parents of children with Down syndrome have misgivings about finding a treatment. The whole disability rights movement more or less followed the civil rights movement. It took years to get children with intellectual disabilities included in regular classrooms. It was a huge effort. But the problem is, the civil rights movement was about people who had a condition, the color of their skin. Down syndrome isn't a condition. It's a medical disorder. We have gotten into this politically correct extreme where some people don't want to admit that. Treating it just means changing the function of an organ, which is the brain, using some kind of medication or whatever means. That's as fair as treating any other disease. I don't see it any differently from treating ADHD. But when you put it all together, it's very hard for a researcher to get funding, because the whole tide is against you."

By the time I finally succeeded in getting him off the telephone, I knew I had found a story that needed to be told.

•—•

A half hour from his office and laboratory at the University of Colorado Denver School of Medicine, where he was then an associate professor of medicine and neuroscience, Costa pulled into a parking space in front of his modest two-bedroom apartment. The figure of a girl in green dashed toward the car—and then vanished.

"Tyche," Costa called to his daughter, "where'd you go?"

We both stepped out to look for her. Coming around the next car over in the parking lot, a Subaru Forester, I found her standing in front of it waiting to get in. Dressed in a lime-colored shirt and skirt, the bangs of her mahogany hair framed by a hair band, sixteen-year-old Tyche stood just four feet, six inches tall, with a round face, broad nose, and heavy-lidded eyes.

Seeing my puzzled look, Costa explained that the Subaru was his car, too—the one he usually drove with Tyche. He led her to the Toyota we'd arrived in, where she sat down in the backseat. As Costa drove us back to his office, I asked what she thought of her father's work.

"He's the greatest scientist," she said, in a slurred, high-pitched voice. Then she added with a laugh, "And he builds evil machines."

"That's from watching too many cartoons," Costa explained. "Her favorite is *Phineas and Ferb*. Of course there's an evil scientist in it who builds all kinds of machines."

"Like the Smell-inator," she added.

Back at his office, Tyche demonstrated to me what people with Down can be capable of even without medication. On the white-board at the front of the room, her father wrote out an algebra problem for her to solve: $8x^2 - 7 = 505$.

"She's one of only two people with Down syndrome who I've ever known to be capable of doing algebra," said Costa, forty-eight, who shares her tawny-brown complexion. "Normally we give her a problem before she goes to bed. It's basically instead of a bedtime story."

"Instead of a bedtime story, you give her a math problem?" I asked, as she got down to $x = 8$.

"Yeah," he answered, "that's how weird I am."

It turns out that with vigorous education and support, many people with Down do far better than once thought possible. Medical care of heart and other physical ailments associated with the disorder have likewise achieved significant benefits, doubling the average life span from twenty-five to forty-nine, in just the fifteen years between 1983 and 1997.

Still, with an IQ that is typically around 50 points lower than average—with some far lower and others, like Tyche, reaching higher—something more than education alone would plainly be necessary to enable the majority of people with Down syndrome to live independently. For Costa, who hypothesizes that memantine might be able to raise their IQ by 15 points, achieving independence for people with Down syndrome is the primary goal. "At some point, you want your children to have their own life," he said. "It's about independence."

Costa's determination to help Tyche and others exceed expectations, he told me, has roots in his own childhood. Raised in rural Brazil, the children of a police officer and a seamstress, Costa and his siblings became impoverished after their parents divorced and their father sent little to support them. Perhaps inevitably for one who had to struggle hard to rise above his background, he comes across as intense and consumed by his work; he hadn't taken a single vacation since Tyche was born. But, he said, he remains equally devoted to Tyche, spending most of every weekend with her and Daisy.

"She's a great kid," he said. "She has a very strong personality. In many ways she has features of a regular teenager. She doesn't like me to get into her bedroom. She loves pop music and vampires." Her relatively high functioning, he told me, is important to him. "If Tyche were really severely affected, I don't know if I would have had the energy of going on with this business." Then again, he admitted to having paternal feelings toward all forty young adults in his study, whose cognitive abilities vary widely. "At the end of the day," he said, "their parents know someone really cares for their kid. It's not an academic experience for me. It's my life."

•—•

"Woo hoo, bring it on!" cried Betsy Baldwin, twenty-five, at the beginning of a ninety-minute test of her memory and other abilities.

She had not yet received any drug treatment, and Costa needed to assess her pretreatment cognitive abilities. With honey blond hair, brown eyes, and freckles, Baldwin wore a pretty flowered blouse and blue-jean capri pants. On her left ankle was an Ace bandage from a fall she had taken a few days earlier while dancing, something she loves to do.

Sitting across from her was Christa Hutaff-Lee, a neuropsychologist on the verge of earning her doctorate. The test, an essential first phase of Costa's study, had been rigorously designed to tease out the neurologic strengths and deficits of Baldwin's hippocampus. In people, as in mice, the hippocampus is essential not only for memory but also for spatial navigation. Down syndrome has been shown to affect both.

After reading off a list of ten objects with exquisite pronunciation, Hutaff-Lee asked Baldwin to repeat them back to her. While I sat quietly in a corner of the small, windowless room, Baldwin took about a minute to recall four of the words.

"Do you remember any more?" Hutaff-Lee asked warmly.

Baldwin furrowed her brow and scrunched her nose. "Give me a hint," she said.

"Oh, I wish I could give you a hint," Hutaff-Lee replied. "That would be good. But I can't give you a hint."

Baldwin repeated one of the words she had already recollected, then sighed, screwed up her eyes, and said, "I don't know."

"Okay, nice work, high five," said Hutaff-Lee, exchanging a celebratory hand slap. "Okay, Betsy, I'm going to read the list again. Are you ready?" After a second reading, Betsy remembered six of the items. After a third reading, she remembered seven. And so it went, turning to tests of finding a virtual "coin" hidden beneath black boxes on a computer touch screen and of following a simple path around the room that Hutaff-Lee had demonstrated.

Simple as the tests looked to my untrained eye, their targeting of hippocampal function is considered as much of a breakthrough as the drugs that Costa and others are studying.

"A mouse can't navigate a maze if you destroy its hippocampus," Costa explained to me the next morning in his small office on the medical school's newly built campus. "The same is true for a monkey and the same is true for a person. We need it to navigate a map and to remember facts."

While a mouse can't take a verbal test, researchers have devised other methods to assess the strengths of its hippocampus. One is the Morris water maze, which measures the mouse's ability to remember the location of a platform hidden just beneath the surface of the water in a circular pool. For another test, the mouse is placed in an unfamiliar cage and given a mild electric shock; twenty-four hours later, when the mouse is returned to the cage, researchers watch how much time it spends frozen in a wary crouch—the normal mouse response to such a frightful environment. While Ts65Dn mice typically freeze much less often than normal mice, plainly not remembering the shock they received on the first go-round, in Costa's studies those injected with memantine spent as much time in the frozen position as their unimpaired cousins.

After Baldwin completed her test, we went out to the lounge area of the testing laboratory, where her family waited. Baldwin's mother, Cathy, teared up at the thought of Betsy being able to pursue her dream of becoming a fashion designer, even as Betsy herself fairly jumped in her seat at the mention of it.

•—•

A year later, in early 2011, the study still wasn't done—finding enough parents willing to participate was taking longer than Costa had anticipated—and the tide seemed to be turning ever more strongly against him, with newspapers touting new studies of the noninvasive blood tests for diagnosing Down syndrome prenatally. Few of the articles, however, noted the profound unease many medical ethicists, including some who are ardently pro-choice, feel about the tests and how they might lead to a dramatic reduction in the Down syndrome population.

"Even people who are traditionally against abortion are sometimes willing to condone it when the abortion is of a fetus with a disabling trait," said Erik Parens, a bioethicist at the Hastings Center in Garrison, New York. "But it's important to recognize that there is a huge range of genetic disorders. In their own way, a lot of kids with Down syndrome flourish, and so do their families."

Advocates of the new prenatal tests insisted to me that parents will be given news of an affected pregnancy by a trained geneticist who will present the information fairly and fully. "It's a gross oversimplification to assume that these tests are going to lead to the wholesale elimination of Down syndrome births," said Stephen Quake, a professor of bioengineering and applied physics at Stanford and developer of one of the new tests. "My wife's cousin has Down syndrome. We just celebrated his twenty-first birthday. He's a wonderful person. It's not an obvious step that you would terminate an affected pregnancy."

But critics of the tests, including Costa and most other parents of children with Down syndrome, insist that such dispassionate approaches are rarely followed in practice, with many obstetricians and genetic counselors providing unduly negative or misleading information. And follow the money, Costa told me: the disorder turns out to be one of the few that actually receives less in research funds from the National Institutes of Health today than it did a decade ago, falling from a high of $23 million in 2003 to an estimated $20 million in 2013. That's less than the $28 million slated for fragile X syndrome (which affects an estimated 50,000 people, at most one-sixth of the estimated 300,000 to 400,000 who have Down syndrome). According to the Global Down Syndrome Foundation, the disorder is now "the least funded major genetic condition by NIH despite being the most frequent chromosomal disorder (1 in every 691 babies in the US is born with Down syndrome)."

"That is a fact," said Mary Lou Oster-Granite, chief of the Mental Retardation and Developmental Disabilities Branch at the National Institute of Child Health and Human Development (NICHD), when I reached her by telephone. "There has been excitement building over the last several years, particularly as more and more information becomes available on drug targets that have been applied to the mouse." But funding decisions, she told me, are not made in a vacuum, and the lobbying of patient groups does play a role.

On a stiflingly hot day in June, I visited Oster-Granite's boss at the headquarters of NICHD in Bethesda, Maryland. The institute is actually named after the late Eunice Kennedy Shriver, President John F. Kennedy's sister, who devoted her life to advocating for the intellectually disabled and who also founded Special Olympics. A plaque with President Kennedy's picture hung on the wall of the office where the director, Alan Guttmacher, sat down to speak with me.

I asked him why about $3,000 in research dollars is spent by NIH for each of the estimated thirty thousand people in the United States with cystic fibrosis, compared to less than $100 for every person with Down, but his answer was rather vague.

"The number affected is a fair metric to use," Guttmacher said. But, he pointed out, some rare conditions have even lower per capita funding than Down syndrome does. Advocacy groups for disorders like AIDS, autism, and breast cancer have certainly played a role in their gaining increased funding, he said. And perhaps, he speculated, Down suffers from an image problem. "Part of it is that Down syndrome has been around for so long," he said.

I pointed out that Dr. Guttmacher's late uncle, after whom he is named, was a president of Planned Parenthood and a vice president of what was then called the American Eugenics Society; that his own work has been primarily in genetics; and that NICHD continues to spend a great deal of its budget on studying contraception. Wasn't there a conflict of interest, in appearance if not in fact, between the mission of detecting (and, if a parent then chooses, terminating) Down syndrome pregnancies and that of funding research to improve the outcomes of children born with the disorder?

"It's a very fair question, I think," he said. "I think it's a potential conflict of interest, but I don't think the way it's done operationally it is. The folks involved in reproductive work are a completely different gang of people than the folks doing Down research."

That same day, I met with Rep. Cathy McMorris Rodgers (R-Wash.), who cofounded the Congressional Down Syndrome Caucus soon after her son, Cole, was born with the disorder in April 2007. The other three representatives in the caucus also have close family members affected by Down. Yet so far they have little if anything to show for their efforts. Even the $5 million written into a law meant to assure that medical personnel who inform pregnant parents of a Down diagnosis provide accurate, up-to-date information on the disorder was stripped out before it was approved and signed into law by President George W. Bush.

"I find myself wondering how NIH really sets their priorities," she told me. "I'm quite concerned that so many of the researchers in the Down syndrome field have difficulty getting funded. Are you aware of the 2007 plan?" she asked, referring to an NIH roadmap that had been devised to set goals for research into Down syndrome, with little discernible progress so far. "They'll give you a huge presentation, a very detailed presentation, as to how they are implementing that plan." She arched an eyebrow and smiled. "As I learn more about the research and the breakthroughs, it's exciting. My fear—I'll put it this way—my fear is that for some they believe that it's been taken care of through prenatal diagnosis."

Seen in context, NIH's dwindling funding for treatment research appears to be only the latest chapter in the history of U.S. government–sanctioned discrimination against people with Down syndrome. As recently as the late 1960s, that history included not only forced sterilizations but immigration policies prohibiting their entry into the United States, along with anyone else deemed to be "mentally retarded." Siegfried Pueschel, born (ironically enough) in Germany in 1932 under the Nazi regime, found out the hard way about U.S. policies when he sought to enter the country in 1966 with his infant son, Christian. Having obtained his medical degree in Germany, Pueschel spent two years as a pediatric intern in Montreal before being accepted by Harvard University to study at its School of Public Health.

"Everyone coming into the United States at that time had to be seen by a physician," Pueschel told me. "We had to go through an examination at the embassy in Montreal. Chris was just a baby, almost a year old. Somehow I mentioned to the physician that he has Down syndrome. He said, 'Oh, then you cannot come into the country.' There was a law at that time that you could not enter if you had an intellectual disability, because in America everyone is smart, of course. So we were not sure what to do. I had my appointment at Harvard already; they were expecting me. My predicament was made known to Senator Ted Kennedy, who was on the board of directors of Harvard at that time. He first of all made sure we could immigrate legally, and then he also tried to get the law changed."

On Monday, June 7, 1971, in a speech to the American Committee on Italian Migration, Kennedy called for immigration law changes to, among other things, "facilitate admission of mentally retarded children whose family members have been cleared for immigration." But to this day, U.S. law still forbids such children from immigrating, although a waiver is available if a parent or spouse would suffer extreme hardship.

Pueschel went on to become the director of the first academic program devoted to the care of people with Down syndrome, at Harvard's Boston Children's Hospital, before taking over as head of the Child Development Center at Rhode Island Hospital in 1975.

"There has been during my lifetime, and since Chris was born, tremendous changes and improvements for people with Down syndrome. When my son was born, on July 25, 1965, I was told, 'Put him away in an institution, he will never do anything, he will be a vegetable, he will be a menace to society,' and all this kind of garbage. This is also what doctor friends of mine recommended. My wife and I said no, he's our son, and we will take care of him. That's the best decision we ever made, because he gave us so much."

The continuing disparity in funding for Down syndrome treatment research would seem to suggest that the more things change, the more they stay the same. Costa, at least, was lucky to be based in Denver, near the home of Michelle Sie Whitten. Her wealthy father, John J. Sie, founded the Starz cable network. After her daughter, Sophia, was born with the disorder, she formed the Global Down Syndrome Foundation and played a central role in establishing the new Linda Crnic Institute for Down Syndrome at the University of Colorado School of Medicine.

Sitting one evening at a posh bar-restaurant in Denver's trendy Cherry Creek neighborhood, sipping prosecco from a champagne flute, Whitten told me how she'd gotten Quincy Jones to be her group's "international spokesman" at a recent ball at which $10 million had been raised. If anybody could kick butt and take names on behalf of Down syndrome, I thought, it is Michelle Sie Whitten. Yet when I posed the hypothetical question of whether she would give a drug to Sophia if it could "cure" the disorder, she bristled.

"The word 'cure' is too charged," she said. "What does that even mean? Your child's your child. I think the idea that you're going to change the child is something abhorrent to parents. For me, Sophia's my daughter, she's fabulous, she has a funny sense of humor, and she just happens to have Down syndrome."

A 2009 survey conducted in Canada asked parents of people with Down syndrome the same question I had posed to Whitten. A surprising 27 percent said they would definitely not give a pill that could "cure" the disorder to their child, and another 32 percent said they were unsure.

Behind the ambivalence toward treatments, parents told me, is a fear that increasing their children's intelligence might change their very identities.

"Nobody would be against giving insulin for diabetes," said Michael Bérubé, author of the 1996 book *Life As We Know It*, published five years after his second son, Jamie, was born with the disorder. "But Down syndrome isn't diabetes, or smallpox, or cholera. It's milder and more variable and more complicated. I'd be very leery of messing with the attributes Jamie has. He's pretty fabulous. At the same time," added Bérubé, director of the Institute for the Arts and Humanities at Pennsylvania State University, "I'm not doctrinaire. If you're talking about a medication that allows people to function in society and hold jobs, how can you be against that?"

•—•

On Friday, July 20, 2012, Costa was scheduled to announce his study's findings in public for the first time. That afternoon, hundreds of people with Down syndrome and their families walked the halls of the Marriott Wardman Park Hotel in Washington, D.C., for the fortieth annual meeting of National Down Syndrome Congress, the largest gathering of its kind.

As I walked toward the room where Costa and other researchers and physicians would be presenting their papers, a young man walking past surprised me by saying, "How are you today?" Looking at his smiling face, I could see he had Down syndrome.

"Good," I said. "How about you?"

"I'm doing good," he said and walked on.

Inside the room, Jamie Edgin, a developmental psychologist at the University of Arizona in Tucson, spoke from the podium. "A lot of us are well aware of progress we've seen in the treatment of Down syndrome that have occurred in the past five to ten years," she said. "There's a lot of excitement."

When it came time for Costa's presentation, he spent more time discussing the drawbacks of his study than its remarkable findings. Released online by the journal *Translational Psychiatry* in coordination with the meeting, the study found that after sixteen weeks, most of the people who received memantine performed slightly better than they had at the beginning of the study on tests of memory. But the effect was statistically significant on only one of the fourteen tests.

"It was a small improvement that was significant on a single measure," Costa said. "But it's the first time in this business anybody saw anything. You can see it as a little study that had a little tiny effect, or as one of the greatest findings in Down syndrome over the past ten years. Both are true."

Then came a presentation from another researcher following the path that Costa had forged. Julie Hoover-Fong, a clinical geneticist at Johns Hopkins University, described a study she was leading of a drug known only as RG1662. Not yet approved for any use by the FDA, the drug is owned by Roche, the pharmaceutical giant. The company was paying for Hoover-Fong to run a randomized trial of the drug, compared to placebo, on thirty-three young adults with Down syndrome. The primary goal of the small study was to demonstrate safety, but it would also look for any sign that the drug might be improving memory.

"I'm limited somewhat by what we can present at this point," Hoover-Fong said. With only the first eleven participants in the study having completed their testing so far, no sign of dangerous side effects had been seen, and there appeared to be a hint of possible benefit. But the full study will have to be completed before any certain conclusions can be reached, she said.

After the presentations, Jamie Edgin told me that parents' concerns for their children with Down syndrome has less to do with IQ per se and more about their activities of daily living. "Many parents say that more language abilities and independence is what they want for their kids with Down syndrome," she said.

George Capone, medical director of the Down Syndrome Clinic at Kennedy Krieger Institute in Baltimore, and co-organizer of the meeting, said he had mixed feelings about drugs to improve cognitive abilities of people with the disorder.

"I don't like to talk about it too enthusiastically," Capone said. "I worry about the way the topic sometimes is promoted. This is very difficult science to do, but it's also very important. I do believe as a clinician, as someone who sees kids and adults with Down syndrome on an almost daily basis, that I would love to have something for cognitive enhancement in my armamentarium. I would not hesitate to say that. I cannot wait till we get to that point."

For his part, Costa said he longs to begin working on a follow-up study of memantine, but did not yet have the funds to do so.

"We need to run bigger trials," he said. "But the biological plausibility is there. We have some clinical data now. We just need to do a bigger trial."

By 2013, two new clinical trials were testing experimental drugs developed by pharmaceutical companies; genetics researchers had succeeded in switching off the third, unnessary chromosome 21 in a cell model of Down syndrome; and researchers like Mobley were telling me that NICHD had greatly increased its focus on finding treatments. As for Costa, he moved his research program to Case Western Reserve School of Medicine, won the National Down Syndrome Congress's Christian Pueschel Memorial Research Award, and was planning to launch a second, larger study of memantine, this time involving two hundred people with Down syndrome.

I found myself wishing it were possible to speak with the author of *Flowers for Algernon*. I wondered what Daniel Keyes would say about how science was finally catching up with his imagination. It would be like interviewing Jules Verne on the eve of the Apollo moon landing. And then I discovered that Keyes is alive and well, living in Boca Raton, Florida, and working eight to ten hours a day on his next novel.

"I always knew this was going to happen, and that there would be a controversy over the idea," Keyes told me when I reached him by telephone (after first correcting me on the number of copies of *Flowers* sold so far: over 7 million, he said good-naturedly, not a measly 5 million). "I think it's a wonderful thing," he said. "Talking about the Jules Verne analogy, is it right to be frightened to go to the moon, to try to explore space? There may be failures in the space program, but I don't think it's wrong to keep trying. The important thing is to never stop trying to improve whatever you want to improve. I've always wanted to be more intelligent than I was. I've always been just a normal guy; I don't have a genius IQ. I've always thought, I wish I could learn everything, know everything."

But didn't things end badly for Charlie, the novel's protagonist (spelled *Charly* in the 1968 movie), as he lost the increased intelligence initially gained?

"*Flowers* doesn't end badly," Keyes insisted. "What these well-meaning scientists wanted to do, they didn't get it right. But I don't think it was wrong of them to try. Charlie didn't become a monster. I always felt that anything a human being wants to achieve is achievable, and the worst thing is not to try."

CHAPTER 10

Clash of the Titans

By the autumn of 2012, Engle and his fellow skeptics were far from satisfied that Jaeggi, Buschkuehl, and others had proved beyond a reasonable doubt that training could improve fluid intelligence. To the contrary, Engle and Hambrick kept telling me and anyone else who would listen that the evidence had already accumulated overwhelmingly in favor of their doubts. So many training studies had piled up, in fact, that two of Randy Engle's fellow skeptics, Charles Hulme and Monica Melby-Lervåg (whose paper on Cogmed was described in chapter 8), had published a new meta-analysis of twenty-three previously published studies of working memory. To some, their results spelled doom for the training field. But by my count, they failed to include at least a dozen other published studies, all of which had found positive results. And even taken on their own merits, the paper's results included many surprisingly upbeat findings. Still, you wouldn't know that from how the paper was spun by some in the press.

The first conclusion of Hulme and Melby-Lervåg's paper was that training produced "large immediate gains on measures of verbal working memory," as well as "moderately sized immediate gains on measures of visuospatial working memory." That sounds good, right? They checked further to look for long-term follow-up results, finding evidence of "moderate" effects on visuospatial tasks up to nine months after training. For "far transfer" to nonverbal reasoning, the kind measured by the gold-standard Raven's progressive matrices, they found an immediate effect that was "small but reliable" in twenty-two studies. And for far transfer to a well-established test of executive attention, the Stroop task, the meta-analysis found ten of the studies showing an effect that was "small to moderate."

All those findings are precisely the kinds of effects once considered impossible, now conceded to have been proved beyond a reasonable doubt by the field's arch-skeptics. But Hulme and Melby-Lervåg brushed past them to emphasize what the young field had yet to prove to their satisfaction: that working-memory training produces long-term improvements in reading, math, and other real-world outcomes.

To Engle and Hambrick, the meta-analysis proved what they had been arguing all along, that working-memory training simply doesn't work, that it's an unethical scam. That dark view would be epitomized by no less a cultural arbiter than *The New Yorker*, in a short blog item posted online by Pulitzer Prize–winning journalist Gareth Cook, under the headline "Brain Games Are Bogus." According to Cook, the meta-analysis flatly concluded that "brain training" doesn't work. "Playing the games makes you better at the games," he wrote, "but not at anything anyone might care about in real life."

Hold everything. Was Cook's 1,715-word verdict on the entire field of cognitive training, coming less than five years after Jaeggi and Buschkuehl's first study—even as hundreds of scientists around the world were cranking out dozens more—essentially accurate? Are all computerized cognitive tasks really just a bunch of nonsense? Had I, the skeptical bastard, been lulled into believing that something as profound and fundamental as intelligence, along with any meaningful real-world abilities, could be significantly improved through minutes-a-day cognitive training?

Final answers are not the business of science, or of journalism, for that matter. But declaring that a still-young field is "bogus" just because a small circle of skeptics continue to have their doubts is way harsh. Let's remember: by the time of the meta-analysis's publication, working-memory training had been shown to improve brain function and structure in multiple fMRI studies, and it had also been shown to work in animals—in mice, for crying out loud. It had been shown to help young children, college students, middle-aged adults, elderly adults, people recovering from chemotherapy and traumatic brain injury, and people with Down syndrome and schizophrenia. The training effect, moreover, had repeatedly been enhanced by transcranial direct-current stimulation. Indeed, a second meta-analysis, published around the same time by researchers at the University of California, Los Angeles, looked at the effects of either cognitive or aerobic training in healthy older adults. After analyzing forty-two studies with 3,781 participants, they concluded that both approaches offer significant cognitive benefits. What's more, two *other* systematic reviews of the literature, both involving older adults, likewise concluded that cognitive training produces measurable benefits. If this is "bogus," what does "effective" look like?

The conclusion I drew from Hulme and Melby-Lervåg's review was that the field of cognitive training remains young, that long-term studies take time (duh), and that more studies are needed. Thankfully, though, that's exactly what the autumn of 2012 and spring of 2013 promised to deliver, as psychologists and neuroscientists prepared for four major scientific meetings, each one successively smaller, more intense, and more telling, where the titans of the field would clash and seek to resolve their differences.

At the first and largest meeting, the annual gathering of the Society for Neuroscience, drawing tens of thousands of scientists to New Orleans in mid-October, ten new studies demonstrated positive benefits of cognitive training. Two were from Sophia Vinogradov, the UCSF scientist who has been testing the computerized tasks developed by Michael Merzenich's Posit Science among people with schizophrenia. Using randomized control groups and running for up to sixteen weeks, both of her new studies showed significant cognitive benefits in the trained groups, with fMRI brain scans detecting normalization of neural functioning, and blood tests finding improved levels of the neurotransmitter D-serine. Two other studies, presented by Mike Scanlon and others at Lumosity, found that training on Lumosity transferred to general cognitive improvement and that peak performance on the tasks persisted for weeks after training ended. (The degree of persistence, however, was stronger in younger adults than in older ones.)

The most impressive training study of the meeting came from Ulman Lindenberger, a professor at the Max Planck Institute for Human Development in Berlin. Even Engle professed admiration for a study that Lindenberger published in 2010, the COGITO study, in which multiple measures of perceptual speed, working memory, episodic memory, and reasoning all showed significant improvements after one hundred days of cognitive training. For the new paper, Lindenberger described the effect of training on the size of different brain regions. He enrolled thirty-five young adults under the age of thirty-two, and thirty-seven older adults between sixty-five and eighty. All were first given MRI brain scans, and half were then randomized to undergo one hundred days of intensive cognitive training. Afterward, all seventy-two participants came back for another round of brain scans. The finding: over the course of the study, among young and old, trained and untrained, a tiny but detectable shrinkage was seen in four brain regions, including the cerebellum and prefrontal cortex. But in those who trained, the cerebellum shrank less. "Think and Grow Brain"? Not quite. It's more like "Think and Slow Brain Shrinkage."

But the Neuroscience meeting was really just a warm-up for the next conference. Drawing about eighteen hundred experimental psychologists (about one-twentieth the size of the Neuroscience meeting), the annual gathering of the Psychonomic Society was where Engle had once given a memorably harsh critique of Jaeggi's work. This year, both were scheduled to return. Arriving in Minneapolis on Thursday, November 15, I headed over to the opening evening's poster session at the city's convention center, where scientists mingled, holding plastic cups of wine. Study authors stood before large printouts (called "posters") of their findings, thumb-tacked to corkboards, giving their raps and answering questions from other researchers who walked by. Meredith Minear of the College of Idaho presented a study of N-back, complex verbal span, and video-game training. She found that N-back training led to significantly better performance on a test of attention and, most interestingly, on the Cattell culture fair intelligence test—a measure of fluid intelligence. Lauren Richmond of Temple University also presented a study showing that transcranial direct-current stimulation significantly improved performance on a complex working-memory task.

Friday morning at noon, I found Jaeggi standing in front of her poster presentation, her long, straight brown hair cascading down over a gray jean sport coat. The study, "Dose-Response Relationship of Working Memory Training and Improvements in Fluid Intelligence: A Randomized Controlled Study in Old Adults," involved sixty-five healthy adults whose average age was sixty-eight. They were divided into three groups: those who did verbal N-back training twice a week for five weeks, those who did it four times a week, and those who served as controls, with no assigned activity. On two measures of working memory, the twice-a-week group scored significantly better than the controls after their five weeks of training, and they showed a trend toward improvement on two measures of fluid intelligence. The four-times-a-week group, however, gained almost twice as much as the two-times-a-week group did on the fluid intelligence measures, reaching statistical significance over both them and the control group.

"I know Randy remains totally skeptical," I said to her, "but it seems like with this study, you've solved a lot of the problems he raised. You've got multiple measures and a dose-response effect."

"You could of course say we could have added an active control," she replied, imagining the critique that Engle might offer. "But the dose effect, that's sort of our control. For this study, it wasn't possible to add another control group as our funds were a little restricted. Also here we couldn't administer long-term follow-up. So we don't know how long these effects will last and whether or not they will go back to baseline."

"It seems like studies are piling up that say yes, they are seeing confirmation of significant benefits," I said. "But then those studies from Randy's group, and the Chooi paper, found nothing."

"It's delicate. It's not an effect you get with whatever you do. We're still at the beginning of a new field."

I wanted to know what Engle made of her new study, so I went looking for him among the hundreds of scientists in the room. Near a guy whose hair had been dyed bright yellow and cut into a Mohawk, Randy stood looking at another poster, wearing a gray sport coat over a red-and-white-checked shirt. I asked if he would like to come see Jaeggi's poster, and he agreed.

Almost as soon as she began explaining it, he interrupted her.

"So this is a passive control?" he asked, raising exactly the concern she had expected about the lack of an "active" control group, the kind that would be assigned to train on something with no likelihood of benefit but would still give participants the impression they were doing something useful and so might induce a placebo effect.

"Passive control," she said.

"And everyone is trained at home?"

"Everyone trained at home. We had a lock file so we could verify whether or not they followed the instructions. I was kind of worried about that. But we were surprised—everyone was following these instructions."

Engle's shoulders heaved as he let loose a big sigh. "So you don't have an active control?"

"In this study, we don't. But here we were interested in—"

In a loud voice, he interrupted her. "Studies must have active control groups, Susanne. To have no control group—this is a no-control group in my opinion."

"Yes, we were interested in the effects of training frequency and seeing whether we would find any dose-response effects."

"Yep," he said, trying to be polite, but not succeeding. "Yeah."

"So both groups improved on training."

"Yep. Right. Okay."

Pointing to one of the graphs showing how much the participants improved on the N-back itself during their training, she continued, "These two, they level out after a couple of sessions and—"

He interrupted again. "Can you give me some idea what the scale is?" he asked, pointing at the same graph. "It's hard to know how big those changes are."

"When you look at the average N-back improvement we saw in our other studies with young adults, they typically improved by about two N-back levels. For this group here it's about one N-back level. It's basically half as much."

"Let me ask you a logical question before you give me the results," he said. "You have argued that in fact the N-back and the complex span task reflect different variance in fluid intelligence. I've got a set of data I'm going to show in my talk tomorrow that, depending on the population, the N-back may account for little variance in fluid intelligence. So when you work on the N-back, what aspect of fluid intelligence are you working on?"

The dynamic had plainly shifted from conversation to interrogation.

"So not from this study," Jaeggi replied, "but from other studies that we did with young adults, it looks like it's a matter of working-memory capacity, in terms of how many items people can store in mind, the resolution of interference or attentional control."

"But that's a leap you're making. I mean, I think interference is probably really important, but you don't have anything directly showing that."

"Yes, I have data that actually speaks to that. Not here. But when you look at some of the transfer measures that we have, the kids get better at interference resolution in N-back. So they make fewer errors in the word trials after training. Only the N-back group does, not the active control group. And also they make fewer errors in the go/no-go trials. Now, for the transfer measures," she said, pointing to the graph showing improvement on working memory and fluid intelligence, "we used composite scores, so combining the matrix reasoning and the block design, as well as the digit span and the letter-number sequencing."

In a tone of incredulity, as though she had committed some terrible faux pas, Engle pointed to the poster and said, "You put matrix reasoning together with digit span?"

"No," she said, correcting him, "the block design with the matrix reasoning, and the digit span with letter-number sequencing."

Pointing first to one group and then the other, he said, "These two are entirely different constructs than those two there. These two are like mixing apples and rhinoceroses."

"Well, we did a correlation and a factor analysis, and they seemed to go together."

Frowning like a judge who just heard a murder defendant explain that the victim had jumped onto his knife, Engle said, "And that's a factor analysis?"

"Well, this is the correlation. We also looked at them separately. But it gives you the same results."

"Yeah, yeah," he said.

"So here," she said, pointing to the second "results" graph, "is the correlation of the working-memory scores. So we had an overall effect that—"

"And the working-memory measure is digit span?" (The digit span task measures how many numbers listed in a row a person can accurately recite backward.)

"Digit span and the letter-number sequencing."

"Do you have it just with letter-number sequencing? Because that's the only thing that's working memory. Digit span is primary memory."

"We see the same results, whether we combine it or not."

"Yep," Engle said, unconsciously shaking his head "no."

"So regardless of how long you train, you improve your working memory compared to no training," Jaeggi said.

"And if you look at just letter-number sequencing, you find the same thing?"

"Yes," Jaeggi said with a firm nod of her head. "Then for the two reasoning measures, now here we see this dose-response effect. So overall training is still better than no training, but—"

"And reasoning measures are what now?"

"Reasoning is block design and matrices."

"But these are two spatial tasks," he said. "To call that reasoning is a stretch. Beyond belief. Quite literally beyond belief. That—that's . . . That—that's . . ." His arm stretched out toward her results, the wrist rotating: palm up, palm down.

Taking a sip from a plastic water bottle she'd been holding, Jaeggi said, "All right, call it—"

"Spatial," he said. "I call it spatial tasks."

"So we call it spatial tasks," she agreed, even though, as has been pointed out many times in this book, the field of cognitive psychology has long regarded matrix tests like the Raven's as the single best measure of fluid intelligence. "So the results on the two spatial tasks, this time the high-frequency training group, in respect to the—"

"All right," Engle muttered. "Cool."

He faintly lowered his head, like a Japanese bowing, then turned and walked away.

And that was it. Watching him go, I felt guilty that I had asked him over, as if I had set up Jaeggi for an ambush.

•—•

The next morning, just before lunch, Engle had his opportunity to present his own new study. As the final speaker on a panel about working memory, he wore a sweater vest over a blue button-down shirt, his sleeves rolled up. He began by recalling his presentation at the same meeting the previous year, where he had, as he put it, offered "a rather intense critique" of Jaeggi's 2008 study. Now, having concluded that N-back might not be an ideal way to train working memory, he described his group's attempt to use complex span tasks for training. "Certainly before you discount that training doesn't work at all," he said, "we thought we should try this."

He'd enrolled fifty-five people and broken them into three groups. The first group did "complex span" training of their working memory for twenty sessions over a four-week period—having to remember a series of letters interspersed with easy math problems. A second group did "simple span" training, which is not supposed to affect working memory—reciting a similar series of letters but without the math problems thrown in as a distraction. And a third group, the active controls, did something called "visual search," which is likewise not supposed to improve working memory. His finding: after twenty sessions, unlike the other two groups, those who trained on a complex span task significantly improved their performance on two other, untrained complex-span tasks, providing evidence of something called "near transfer" to a general improvement in working memory. But for "far transfer" to improvement on measures of fluid intelligence, Engle said, "It's a very simple story. No effect." But Engle had also tested for something he called "moderate transfer," on a test known as the "running letter span," where a list of letters is spoken aloud and then suddenly stopped and a person is asked to remember as many back as possible. On the running letter span, he said, "We saw a rather sizable effect, but it's hard to know what that means. I'm not sure how to interpret this." Likewise, he noted, on a test of secondary memory, he saw "a rather sizable effect."

His conclusion, he said, was that "training on complex span tasks shows transfer to other complex span tasks. But we see no evidence for transfer to fluid intelligence."

Then it was time for questions from the audience. Given Engle's many previous studies demonstrating how closely related working memory and fluid intelligence are, the first question seemed inevitable: "How do you reconcile the fact that training improved complex span by three standard deviations, yet general fluid intelligence didn't change at all?"

"Let me point out," Engle answered, "that the relationship between height and weight is almost exactly as close as the relationship between working memory and fluid intelligence. But would you argue that if you make me fatter I'll get taller? I've gotten a lot fatter since I was twenty-one; I haven't gained an inch in height. We experimental psychologists have a hell of a hard time parsing these correlations. But working memory and fluid intelligence are not the same thing. These are severable constructs."

The next question was a follow-up: "So what do you think you *are* training that is showing up in the span task but is not showing up in the fluid intelligence measure?"

"You know, well, one possibility, and I don't know," he began, "I don't have studies to show this, but one possibility is you're learning how to actually juggle two things better. You're learning how to block distraction. And I think proactive interference is a really important factor in all of these things."

For the final question, a young woman asked about what he made of a study showing that working-memory training improves reading comprehension.

"Are you talking about the Jason Chein study?" Engle asked, and the woman said she was. "I have no comment on that. I mean, is Jason here? I don't want to offend him if he's not here. If he was here I would offend him to his face. You look at that study, you have to look hard to find any effect. Maybe it's there, maybe it's not. People do funny things when they're analyzing these data in weird ways. It bothers me. I never review them. But I'd like to see that for real. For real."

Six hours after Engle's talk, the final poster session of the meeting was held, on Saturday evening. Citing Jason Chein's paper as an inspiration, researchers from the University of Pittsburgh's Center for the Neural Basis of Cognition described their carefully constructed study of forty-five English speakers who were given either adaptive working-memory training, which increased in difficulty as they improved performance, or nonadaptive training. In a startling demonstration of far transfer to a totally untrained (and potentially highly useful) task, those who trained adaptively showed gains in working memory that correlated with an improved ability to learn Arabic vocabulary words. "These results suggest," the study concluded, "that improvement in working memory capacity, through adaptive cognitive training, transfer[s] to second-language learning."

•—•

As the Psychonomic conference had been less than one-twentieth the size of the Neuroscience meeting, so the International Society for Intelligence Research conference was less than one-twentieth the size of Psychonomic. I counted just seventy-five scientists in attendance for the three-day meeting in San Antonio, held in mid-December. Engle didn't attend, but Jaeggi was scheduled to cohost a symposium on improving intelligence, where many senior researchers would have the opportunity to grill her.

My first impression upon looking around at the meeting room was that the group could just as well call itself the International Society of Old White Guys in Blue Sport Coats. Strictly speaking, they were not all men; fourteen women showed up. And they weren't all old: some were in their forties; a few in their thirties. But they skewed much older than a typical academic meeting, and not one person of color was among them. No doubt that was due to the racist reputation of some of its members, two of whom had died during the preceding year: Phil Rushton (whom Engle had decried as "the worst" when we spoke) and Arthur Jensen. Rushton had long been dismissed by many as a crackpot and creep, the kind who handed out forms to his students asking them to list their penis size and sexual habits. Jensen, on the other hand, was considered by some to have been unfairly villainized after he published a paper in the *Harvard Educational Review* back in 1969, "How Much Can We Boost IQ and Scholastic Achievement?" The extraordinarily long and carefully researched paper, running 125 pages, had summed up the dismal view (one that held sway among many academics until publication of Jaeggi and Buschkuehl's paper nearly forty years later) that genes account for most of a person's intelligence, and for most of the average differences between the races, and that therefore not much can be done to make a difference.

Still, the meeting had drawn not only Jaeggi but also Adam Russell of the Intelligence Advanced Research Projects Activity (IARPA), the government's version of DARPA for the spy community. He was seeking grant proposals for a new program called Strengthening Human Adaptive Reasoning and Problem-Solving (SHARP). If they considered the conference worth attending, I figured I was in good company. And I hoped to see how Jaeggi's views stood up against conservative scientists like these. If they believed it, who other than Engle and his clique of skeptics could continue to doubt it?

Ironically, the very day the meeting began—Thursday, December 13—the *New York Times* printed an op-ed column by Nicholas D. Kristof entitled "It's a Smart, Smart, Smart World." The column took note of research by New Zealand scholar James R. Flynn, who had been among the first to observe that average IQ scores around the world have been steadily rising for a hundred years—so much so that a person who scored 100 back in 1900 would now, according to our current averages, receive a score of under 70, low enough to be considered intellectually disabled. This upward arc of IQ scores, known as the Flynn effect, is seen by many as evidence that genes do not limit intellectual abilities and that the social, nutritional, and educational environments play a major role. "The implication," Kristof wrote, "is that there are potential Einsteins now working as subsistence farmers in Congo or dropping out of high school in Mississippi who, with help, could become actual Einsteins."

As it happened, an entire session of the meeting was devoted to analyzing the Flynn effect. Some researchers pointed out that the IQ gap between blacks and whites in the United States has been narrowing for decades, although the gap between rich and poor stopped shrinking late in the twentieth century. "The talented poor are being left behind," said Jonathan Wai of Duke University. "They rely on funding for gifted and talented programs, and we have seen zero increases in such funding."

The meeting's keynote address was delivered by Craig T. Ramey, founder of the widely acclaimed Abecedarian Early Intervention Project. Beginning in 1972, the program delivered five years of intensive, high-quality child care and educational stimulation to fifty-seven North Carolina infants from poor, mostly African American households, and compared their results to fifty-four infants from a similar background who received only nutritional supplements, social services, and health care. By age thirty, those in the intensive treatment group were four times more likely to have graduated from a four-year college than those in the control group, five times less likely to have used public assistance in the previous seven years, significantly less likely to have been involved in crime, and delayed becoming parents by nearly two years. Increases in IQ, however, were only modest: 4.4 points on full-scale IQ. Some researchers considered those results to be proof of how difficult it is to affect IQ, no matter the intensity (and cost) of the program. But in his address, Ramey said that such criticisms are irrelevant, given the huge improvement in life outcomes the program produced.

"What we're doing for poor children now is a pale imitation of what we know to be effective," he told the group. "A program like ours can be run anyplace in the United States for an average of about $11,000 per year, per child. So you say we can't afford it, right? But we can afford to stick 'em in prison when they get older, and we can afford to pay for special education when they're failing in schools. I calculate that the least rate of return for every dollar invested in a program like ours is four dollars. The economic argument is a red herring dragged across the stage by regressive conservatives who can't stand the idea of helping poor kids. If we had taken the same stance on public health in the 1950s, we would still have kids dying of polio in iron lungs."

As sensible as his argument was, and as beneficial as the Abecedarian Project had been, it occurred to me that the chances of politicians approving a program costing $11,000 per year for every poor preschooler in the United States, or any other country, is approximately zero. Of course an enriched environment is good for kids. Who can doubt that? But getting taxpayers to fund such a program is another matter entirely. Which is why cognitive training, if it really works, is so appealing: it's practical, efficient, and inexpensive.

As the meeting proceeded, two researchers working independently of each other, Clayton Stephenson of Claremont Graduate University in California and Edward Necka of Jagiellonian University in Kraków, Poland, each presented a new study showing that working-memory training led to increases in intelligence. Nicolas Langer of Harvard Medical School also presented a careful study showing that working-memory training induced beneficial changes in brain function.

On the morning of the final day of the meeting, Jaeggi presided over the symposium on improving intelligence. The first speaker was Earl "Buzz" Hunt, professor emeritus of psychology at the University of Washington. Hunt reviewed evidence from studies dating from back in the 1980s. Raymond Nickerson of Tufts University and Marilyn Carlson of Arizona State had both shown that teaching critical thinking skills could help children solve complex problems. "One of the strategies," Hunt said, "was that before you try to solve a problem, describe it to yourself. That's a very good strategy in life, and it produced substantial improvement. Another approach tried in the 1980s, and it's almost totally ignored these days, is the Venezuela Intelligence Project. The country tried to increase the intelligence of their children by teaching thinking skills. And what happened: intelligence test scores grew significantly. But that experiment ended when the political control of Venezuela changed."

Venezuela wasn't the only place where the teaching of critical thinking skills was frowned upon, Hunt said. The Republican Party of Texas, he said, had written into its 2012 platform that it opposes "the teaching of Higher Order Thinking Skills . . . critical thinking skills and similar programs that . . . have the purpose of challenging the student's fixed beliefs and undermining parental authority."

"So I'd better get to another topic quickly," Hunt quipped, "before the Texas Rangers come in here and take me off the podium."

The next speaker was Roberto Colom, a psychologist at the Universidad Autónoma de Madrid in Spain. He presented the results of a new study on which he and Jaeggi had collaborated, involving fifty-six adults. Half had been assigned to four weeks of dual N-back training, half to serve as controls. Significant improvements were seen with fMRI in the structural integrity of brain regions associated with intelligence for those who had trained, but not for the controls. And gains in fluid intelligence were also seen in the training group, although they fell a single percentage point short of the standard cutoff used for demonstrating statistical proof.

Jaeggi then took the podium to present an overview of the field since publication of her 2008 paper, reviewing dozens of papers that had replicated hers. "We really think the question at this point should not be whether cognitive training works, but for whom and why," she said. "There are still lots of things we don't know. What are the underlying cognitive mechanisms? What needs to be done to make the effects stronger? Are booster sessions necessary to maintain the effects? And most important of all—this really is close to my heart—to what extent do the improvements affect academic achievement and other real-life outcomes?"

The next speaker was Richard J. Haier, a psychologist at the University of California, Irvine, who has been studying the neural basis of human intelligence for many years. His talk was entitled "Intelligence, Cold Fusion and Dark Matter."

"If the title of that first paper published by Susanne had not included the term 'fluid intelligence,' if she had just stated it a little differently, I don't think there would have been much controversy about it," Haier said. "Independently of Randy Engle, when I first saw the paper, I thought, 'This is like cold fusion.' I thought it couldn't possibly be true. But increasing intelligence is kind of what we're all trying to do, whether we acknowledge it or not. We're clearly on a path toward doing that biologically, with drugs."

As he continued, Haier used the technical term "g," coined decades earlier by psychologists to describe intelligence as a biological construct.

"Let me talk about dark matter," he said. "The study of g is like the study of cosmology. They both involve big questions. In cosmology, one of the great mysteries is the nature of dark matter, because dark matter is inferred by equations. No one's ever seen it. No one's ever measured it. Physicists know it's there. It's just like g, isn't it? We infer it, but we don't have a direct measure of it. All our psychometric tests, every one of them, is an estimate of g, an indirect estimate of g. So to answer the question of whether we can increase g, we really haven't had the tools. The standard psychometric tools were not sufficient to answer that question. But now, with modern brain-imaging technology, we are beginning to have the capability to measure the flow of information around the brain, millisecond by millisecond. I think this is going to be a transition time for the study of intelligence, focused on whether you can increase g. I think the tools to answer that question now exist, and it's really going to be exciting to put all this together."

Following Haier's talk, the floor was opened to questions and comments from the audience. Doug Detterman, one of the older researchers in attendance and the founding editor of the journal *Intelligence*, was the first to speak. He had previously been quoted in news reports expressing grave doubts about training.

"It's not that people don't want it to succeed," he said. "Everyone here would like to find ways to increase intelligence. But the skepticism you hear from us older folks has to do with the number of times our hearts have been broken. There have been a number of cases, probably hundreds of cases, where these kinds of breakthroughs were claimed and they turned out to be not methodologically supported. So it's very important to consider the methodological issues in this research. I would love to have it succeed. I just hope our heart isn't broken again."

Next up was a British researcher, James Thompson, of University

College London, who had previously published research showing that the average IQ of a country is related to its economic output.

"Thank you for showing me the N-back test," he said, his voice suffused with the upper-class accent of a proper English gentleman. "What a dreadful waste of a life! This is not something that anyone with a mind should be involved in. It's a total and absolute waste of time. I'm just on the border of profanity. Look at some of the nonsense we've heard about over the years. Remember sleep learning? People had little earphones they would wear, to learn while they slept. This is part of the dream of getting something for nothing. Forget this. Please."

Colom, holding a portable microphone, asked, "James, do you go to the gym three or four times a week for taking care of your physical fitness?" He gave the microphone back to Thompson.

"This is a personal question," Thompson said. "To a British person this is extraordinary. I have been going to a swimming pool for eight years. It's had no effect on me at all, but I still go."

"I tend to disagree," Detterman said. "I think it's all right to do these things, even if they don't come out so well. Because I think we learn a lot, and we're learning a lot about appropriate methodology."

A balding researcher in the back, wearing glasses and a trim white beard, took the microphone.

"I guess I qualify as an older researcher now," he said, "but I am not skeptical. When you think about this research in the larger context of everything we've learned in neuropsychology about the plasticity of the brain over the past fifteen years, there really ought to be ways to train certain functions if you believe in plasticity, which is really indisputable at this point. Now, the working-memory line of research has been going on for a very short period of time. It's just not reasonable to expect that we would have identified all of the active ingredients in the short period of time we've been studying it. But the early results are, I think, promising. This line of research ought to continue. We ought to continue to actively identify and map the ingredients that are crucial."

Then Detterman asked perhaps the most daring question of the meeting. Addressing Jaeggi, he said, "How do you pronounce your name?"

"My last name?" she said. "YAH-kee."

"And it's Su-ZAN or SU-san?"

"I go by SU-san," she said.

"Thank you," Detterman said. "That's important."

The fourth and smallest conference of all promised to offer a final distillation of the state of the young science of brain training, or at least a final smack-down, with just fifteen leaders in the field, including Jaeggi and Engle, scheduled to present and defend their latest findings during the course of a single day: June 10, 2013. Arranged by Harold Hawkins, the Office of Naval Research program manager who had been the leading funder of their research for the past few years, the meeting included about twenty invited guests, naval officers with military ribbons on their uniforms among them, who came to figure out whether the money being spent would ever pay off in the form of smarter, more capable navy personnel.

"This is a secure facility," Hawkins reminded the group as he convened the meeting precisely at 8:00 a.m. "You should restrict your activities to this room, the hallway, and the bathroom."

He sat at the head of a long mahogany conference table, his fifteen grant recipients arrayed around it, the rest of us sitting against the walls. We were on the eighth floor of a building in Arlington, Virginia, in the wood-lined executive conference room of QinetiQ North America, a private company that had once been part of the

United Kingdom Ministry of Defense, Defense Evaluation and Research Agency.

Although I had spoken with Hawkins many times by telephone, this was the first time we had met. A picture had formed itself in my mind of a broad-shouldered marine type, but Hawkins turned out to be thin and gray haired, with sunken cheeks and red-lidded eyes.

"We're going to hear about a pretty large investment I've been making in something called brain plasticity and cognitive readiness," he said. "My interest in brain plasticity goes back to a literature suggesting that with a very brief period of training, it's possible in young adults to enhance some of their cognitive abilities, their executive control capabilities, maybe even aspects of their intelligence. If this is true it has very profound implications for the military and society at large. I wanted to replicate and extend the findings in the literature. I wanted to understand the mechanisms underlying the neurobiology and cognitive processes. That has led to a pretty major investment from part of my core program, money used at my discretion."

The problem, he said, is the discrepancy among different researchers' results. "In the case of fluid intelligence, some people have gotten some pretty good effects. Other people have gotten no effects whatsoever. That's important, because it's going to define whether this is something that makes a contribution to influence our abilities to affect cognitive capacities or not."

His continued interest, however, comes from the highest levels of the military. "Almost all operational needs statements in the military and navy emphasize cognitive resilience," he said. "That's seen as very desirable by military decision makers."

Even so, Hawkins emphasized the need for clear results that can be turned into reliable training programs for navy personnel. "Over the past ten years," he said, "I've been very fortunate to have quite a number of my research programs transitioned into systems training programs that are being used right now out in the fleet." As an example, he described a simulation program, developed under his support, that now trains sea combat commanders and over fifty support personnel on war scenarios involving the deployment of fast-attack craft, mines, torpedoes, missiles, submarines, and more—all at the same time. "You've got to juggle all these assets," Hawkins said. "Nowhere is there any training for these situations, except in this facility we've developed. It's a pretty big deal. It's something I'm particularly proud of. So my customers are not only people like you, but the fleet. I take that very seriously."

The first scheduled speaker was Art Kramer, the University of Illinois researcher whose studies of cardiovascular exercise as a means of improving cognitive abilities are described in chapter 4. His military research went back over thirty years, he said, when he collaborated on creating Space Fortress, a computerized training program that was pioneering in its time but now looks laughably outdated. Even so, he pointed out, "It's the only game to show transfer to real-world, military piloting." In his latest research, Kramer has taken twenty cognitive training games that are freely available online and is assessing which ones work best for improving particular abilities, with the goal of being able to tailor particular kinds of games to particular needs.

Shortly after 9:00 a.m., Engle stood up from the chair where he had been sitting beside Jaeggi and went around to the front of the room to give his talk. Instead of giving his usual critique of others' research, Engle talked in positive terms about new research he was doing on training attention control. In particular, he was interested in "updating" a person's ability to quickly and repeatedly refresh the object of his or her attention.

"This updating construct is really key," he said. "If we're going to focus on training, we have to focus on that. I think navy recruits and all of us can learn to do it better. Updating is a nice, coherent variable. The secret is to come up with the aspect of updating that is generalizable."

So rather than continuing to fight on the bloody field of working-memory training, Engle had moved to less trampled ground on which he could plant his flag: attention updating training.

But that didn't mean nobody was left to fight the good fight against working-memory training. Immediately following Engle's talk came Mike Dougherty, a psychologist at the University of Maryland, who described two large studies of working-memory training, one involving 121 college-age students, another involving 132.

The first study, he said, appeared to show transfer of training to improvement on untrained tasks. "We were quite honestly optimistic about this finding initially," he said, "because it looked like we found pretty good transfer. The problem was that all of our transfer tasks shared stimulus properties with our training tasks. That led to a light-bulb going off about the true scope of our transfer."

In the second study, he found, "People improved only on the things they trained on. People who trained on N-back improved on that and nothing else. People who trained on spatial tasks improved on other spatial tasks and nothing else. We showed no crossover transfer effects. In fact, on reading, our training group got slightly dumber."

Wearing a bow tie, glasses, and a checked jacket, his hair cut short, Dougherty reminded me of a NASA engineer out of the 1960s. His use of erudite statistical measures seemed very much like that of a rocket scientist, on whose calculations an astronaut's life depends. And according to his analysis of his second study, there was better than four-to-one odds that working-memory training did not transfer to general, untrained abilities.

"Even if you believe a priori that working-memory training works," he said, "you should now, in light of our data, have a slightly more pessimistic view."

Engle was the first to comment. "I don't see that as pessimistic," he said. "I see that as optimistic. We can identify tasks that are quite specific. If you went into one of those training programs Harold showed us at the beginning, where a commander has to decide between launching submarines or boats, each of those processes that go into that multitasking is understandable and discoverable. We can identify and teach each of those."

Sitting against the back wall, Joe Chandler, a young research psychologist with the Naval Medical Research Unit in Dayton, Ohio, raised his hand to speak.

"I'm not convinced that far transfer to untrained abilities is as important as some people think it is," he said. "You can take someone to the gym and teach them to do squats, and those squats won't make them better on curls. But they're still going to do better because their stability is better."

Lou Matzel gave the next presentation, announcing the latest results on his amazing mice. In a new set of experiments, not only did working-memory training on the dual maze improve their general intelligence, so did twelve weeks of exercise training on a running wheel. His most remarkable new finding was that mice who did both working-memory and exercise training increased their general intelligence more than twice as much as those who trained on just one or the other. In fact, the results of doing both simultaneously were greater than the sum of the effects of doing either alone.

"This probably has important implications, given that it's so different from the way we humans normally do things," Matzel said. "Your typical undergraduates spend nine months studying and drinking beer and chasing girls, then they go off for three months during the summer to exercise. It looks like the best results come from combining exercise with learning."

Mike Posner of the University of Oregon, whose research into mindfulness meditation is described in chapter 4, spoke last before the lunch break. Bearing a strong resemblance to the actor Martin Landau, who began his film career in the 1950s, Posner spoke with a booming, vigorous voice.

"According to our theory, mindfulness meditation is comparable to physical activity, in that it changes the brain state," he said. "Working-memory training and meditation involve very different areas of the brain. They seem to have different anatomies. It does seem sensible to ask if there's some kind of synergy to be gained by combining them." Although he hasn't done such a study, he did present his latest results on a randomized trial of sixty adults in Texas who were offered mindfulness meditation. Half of the participants were smokers, and half were nonsmokers. Although they were never told that the study was intended to reduce their smoking—and although none of them were intending to quit—those assigned to the meditation group showed a 60 percent decline in the number of cigarettes they had smoked, based on a breathalyzer measure.

Again, Engle was the first to speak up, in support of the training. "Attention control plays a big part in a long list of psychopathologies, including drinking alcohol," he said. "I would expect a lot of overlap with these kinds of outcomes if you focus on attention control."

"One of the first things we saw from this mindfulness training was on the conflict-resolution measure from the Attention Network Test," Posner said.

"That's where we've shown big differences, too," said Engle.

Who was this man?

After lunch, Ed Vogel, a colleague of Posner's at the University of Oregon, spoke to the meeting via speakerphone. "My wife and I are expecting a new baby any moment," he said, "so you'll have to forgive me if I end my talk abruptly."

As Vogel's fuzzy voice droned on about attention capacity and how it varies in each of us from minute to minute, hour to hour, day to day, depending on our circumstances, I struggled to stay awake and listen. His key finding from a study involving an unusually large number of participants, 495 undergraduate students, was that those with lower levels of working-memory capacity tend to entirely zone out when the number of items they are asked to keep in mind grows too large for them. Although they can remember three items nearly as well as people with higher working-memory capacity, once the number of items reaches four or more, their recollection accuracy nose-dives to just one or two items.

"Really nice work, Ed," said Engle. "You've called attention to something I've found in my lab for the last fifteen years. As tasks become harder and harder, low spans at some point just basically give up." (By "low spans," Engle meant people with low working-memory spans, that is, people whose working memory is less robust.) "I had a paper in 1992 where we had people press a key as span tasks got harder. After high-spans reached their peak, they kept trying harder and harder, but low spans just dropped off like a rock. It's a giving-up thing."

"Your study inspired some of my own," Vogel replied. "In some of our latest experiments, we've been testing a strategy to help the lower-capacity individuals. We give them six items, but instead of telling them to try to remember all six, we tell them to try each time to remember at least three. That seems to increase their average performance. We're not necessarily trying to make the low-capacity individuals smarter, we're just trying to make them dumb less often."

Shortly before 4:00 p.m., John Jonides, the former supervisor of Jaeggi and Buschkuehl at the University of Michigan, presented the results of a new study on which they had collaborated. Since the time I had first met them in Michigan a year and a half earlier, they had been telling me about their belief that paying people to participate in training studies undercuts their performance. They had paid participants little or nothing in their early studies, where the effects of training were greatest, whereas Engle had paid his participants as much as $350 to show up and had found far fewer effects. One possible explanation for Engle's lack of results, they believed, was that the sorts of people who sign up for a study in expectation of receiving payment are not intrinsically motivated and so do not really work at the training program. Jonides' group set out to test their hypothesis: they recruited two groups to the same N-back training study, some with flyers offering $350 to participants who stuck with training till the end, others with flyers offering no money but explicitly promising the possibility of "brain training."

"So here are our results," Jonides said, clicking on a slide. "People who improved more on N-back performance had greater transfer to measures of fluid intelligence. But this was true both for the paid and unpaid people. So our hypothesis was wrong. Paying doesn't make a difference. That surprised us. But now that we know that doesn't hurt, we should have an easier time attracting participants to our studies."

This is what's cool about science: facts win, even when they're disappointing to the scientist. But with Engle now publicly conceding that some kinds of training might work after all, it seemed only fitting that Jonides, Jaeggi, and Buschkuehl eat their own share of crow.

The next-to-last speaker of the day was Jason Chein, the Temple University psychologist whose previous work Engle had maligned at the Psychonomic meeting. Chein presented a new study on a novel task requiring visuospatial working memory. Participants practiced looking at two-dimensional pictures of three-dimensional cubes linked together like building blocks and then had to decide whether a second image was of the exact same group of cubes shown from a different angle, or a different group.

"People got better at the task," Chein said, "but we saw no transfer to another working-memory task. I was hoping it would work out. This was a disappointment."

Caw! Caw! Damn crows.

But Chein had two other studies to present. For this next one, he trained people on something called rapid instructed task learning, or RITL. It's harder to explain than it is to learn, but essentially they seek to exploit people's ability to follow a random set of instructions, as in "Press the L key only if the four words you see on the computer screen have two syllables and describe objects that are inanimate and green, but not if they all start with the same letter."

"We have previously seen a nice transfer of working-memory training to certain measures of cognitive control," Chein said. "Here we get a nice replication. RITL produced a benefit on the Stroop task. In the past, we saw no transfer to a measure of fluid intelligence. Here we have the slightest hint that we are moving fluid intelligence. We saw nothing on the Raven's, but a hint on the Cattell. It wasn't strong enough to be statistically significant. But it's the first time I've seen the needle move at all."

His third and final study combined working-memory training with transcranial direct-current stimulation. After just ten days of training, "We saw a significant effect of training by itself, which became stronger when it was combined with tDCS."

The final speaker of the day had collaborated with Chein on some of his latest studies. Like Engle, Hawkins, Kramer, and Posner, Walt Schneider was another grand old man of the field, although he definitely had the most distinguished, ready-for-prime-time look and had appeared on *60 Minutes* to discuss his fMRI studies of the brain of Temple Grandin, the animal scientist, author, and autism activist. A professor of psychology and neurosurgery at the University of Pittsburgh, he is also the senior scientist at its Learning Research and Development Center. He also happened to be one of the few researchers at the meeting I had never spoken with, so I greatly looked forward to hearing his view.

"I've been sitting here listening to all these talks on working memory," he said. "I want to use the analogy of physical training. The military spends a lot on physical training, and they're very good at it. They can triple a recruit's upper-body strength. That same training also improves a recruit's general health, focus, grit, and discipline. The benefits are easy to achieve at low cost. Now, in terms of working memory, we have also seen greater transfer for more intense effort. I take this as a sign we're succeeding at component training. I think this is a case where the cup is more than half full. I would never have believed we could change working memory with the task Jason just described. In 1977 I said working-memory capacity doesn't change. But then again, if you had a back problem, from 1950 to 1999, your doctor would have said to stay in bed for three days of total rest. Now they say get up and move. So the science is moving forward.

"Now, starting this summer, I will be bringing in a group of twelve military personnel who have experienced traumatic brain injuries. I need to know which tasks work best for which kinds of deficits. Some of them have given up two standard deviations of IQ for their country. They went from 120 to 90, from a manager to a clerk. I need to know how to assess them and exactly which tasks are best to help them. I want your help. I've got to do this. We don't have the luxury of waiting twenty years."

There ended not only the meeting but my story of the birth of a new science. Judge for yourself, but from my vantage point, scientists like Jaeggi, Jonides, Buschkuehl, Posner, Kramer, Merzenich, and Klingberg had overturned the orthodoxy that the brain we have is the brain we're stuck with, that improvement is impossible and getting smarter is a pipe dream. Engle had argued that his handful of studies showing no effects was the beginning of the end. Instead, while he remained as harsh a judge of some researchers' work as ever, his conversion to the group of believers in the possibility of training—even if only his own particular version of attention training—marked, from where I sat, the end of the beginning.

There remains much work to do. No one can yet tell Walt Schneider exactly how best to measure his wounded warriors' deficits nor which of the many training programs being studied and sold would be best to treat them. I can't tell you whether to try Lumosity or LearningRx, meditation or exercise, learning a musical instrument or getting your brain zapped by tDCS. I can confidently say, however, that combining exercise with a purely cognitive training method is likely to give better results than either alone; that doing *something* new and extremely challenging, rather than sticking to your old ways, is likely to pay off in innumerable ways; and that the weight of the scientific evidence strongly suggests that training your working memory and attention is likely to improve your general mental abilities and your ability to learn.

Imagine if, instead of wasting time playing solitaire on their smartphones, all those people we see on buses, trains, and planes were making themselves smarter by playing working-memory games.

Imagine if the nearly one in ten teenage boys now taking stimulant medication for ADHD had a nondrug option for increasing their attention and ability to learn. Imagine if middle-aged workers and retirees could regain their youthful cognitive speeds, and if poor, inner-city kids could get an intellectual lift with a straightforward training regimen at modest cost.

These are no longer fantasies. Despite what a century of skeptics have claimed, brain training is not bogus: scientists have tested it and argued over it and reached a consensus that you really can make yourself, your child, or your parent smarter.

Which, come to think of it, is what I tried to do.

CHAPTER 11

Final Exam

My interest in intelligence goes back further than I have let on. When I was in third grade, I couldn't read. I can remember sitting in Mrs. Browning's third-grade classroom in Whittier School, when she came over to my desk and tried to get me to read a few sentences from a Dick and Jane book. She pointed to a word and asked me to pronounce it.

"Tuh-hee," I said.

"*The*," she said, correcting me, and that's when it clicked—the moment when I learned to read the word "the."

Growing up in Teaneck, New Jersey, in the 1960s, I was what Mrs. Browning politely called "slow." During the parent-teacher meeting, she actually told my mother, "Daniel is a slow learner." I sat during lunch in the gymnasium with the—forgive the term—dumb kids. I was grouped with them during reading and math: the "slow group."

And then, in fourth grade, I was rescued by *Spider-Man*. My best friend, Dan Feigelson, who lived down the block and was reading chapter books by kindergarten, had started reading *Spider-Man* and other Marvel comics with some other kid, and together they began drawing and writing their own comics. In response to this loathsome intruder's kidnapping of my best (and, I confess, only) friend, I began reading comics, too, and then began scrawling and scribbling my own. Soon Dan Feigelson and I were happily spending every afternoon on our masterworks, while the odious interloper was never heard from again. We even convinced Dan's father, Dr. Feigelson (rest his soul), to shoot a super-8 movie that we scripted, "Bob Cat and Bat vs. Disappearo!"

By sixth grade, I was getting straight A's.

So what happened there? Was Mrs. Browning right—was I actually "slow" in third grade, and did I then somehow literally become smarter because of immersing myself in reading and writing comic books? Did *Spider-Man* do for me what chess had done for Julie Vizcaino?

•—•

On Sunday, January 13, 2013, I walked through the frigid cold into a Minneapolis library to retake the Mensa IQ test. My original hope of improving enough from my three and a half months of training to qualify for membership had already been quashed, back in November, when I learned that I already did qualify based on my first test. To my astonishment, my IQ, according to the first test, was 136, putting me in the top 2 percent of people. (Interestingly enough, that was almost identical to the IQ score that could be calculated on the basis of my combined SAT scores, from back in 1975.) Now all I wanted to see was whether that first test score was a fluke, or whether I would actually raise it.

A few weeks later, back home in New Jersey, an envelope arrived. Ripping it open, I saw that on the basis of the Mensa admission test, my IQ score had gone up—by one stinking point, from 136 to 137.

By some estimates, this was enough to place me in the top 1 percent of scores. But I was disappointed. Didn't that prove that brain training doesn't work, after everything I'd done?

Next, I flew back to St. Louis to have Mike Cole run another fMRI scan of my brain and to fight the fear of being strapped down in that machine again. Weeks later, he e-mailed me to report that he had found no change in the connectivity of my prefrontal cortex to the rest of my brain. It was especially disappointing because in the first week of January, when I had already given up hope of ever getting my brain zapped with tDCS, Felipe Fregni, an associate professor of neurology at Harvard and director of the Laboratory of Neuromodulation at Spaulding Rehabilitation Hospital, had finally agreed to treat me with it for three twenty-minute sessions on three consecutive days. Just as expected, it had felt like little more than a tingling where the applicators were applied to my scalp. And on the second day, my performance on the N-back actually jumped to its highest ever. Five days is the least amount of exposure he and other researchers have tested, though, and ten days has generally been considered the most effective. But Fregni and I could find only three days before my follow-up Mensa test when he and I were both available for the treatment, so, as he predicted, the lasting effect was obviously negligible.

Even so, sitting in the cafeteria of the rehab hospital after my third tDCS treatment, I did have the opportunity to reflect on the distinction between "rehabilitation" and "self-improvement." A young man I'd seen on the seventh floor that afternoon was being trained to use robotic legs. "Step," the trainer next to him had kept saying. "Step." What was it I was attempting to achieve that was so different from what the "disabled" man was attempting? Aren't we all balanced between worse and better, standing at the fifty-yard line of our selves, no matter how we compare to others?

All that awaited me were my follow-up tests of fluid intelligence to be administered by Buschkuehl down in Maryland, on Tuesday, January 15, and Wednesday, January 16. Driving down the New Jersey Turnpike, I considered all that I had been through.

The mindfulness meditation had been a washout. Hey, there's only so much time in a person's life.

Learning to play the lute, on the other hand, had amazed me. After just three months, I could read tablature to learn new pieces on my own and play a few songs that sounded, to my ears, simply beautiful. I watched the fingers on my two hands move as if they belonged to somebody else. I honestly didn't believe how much progress I had made and how lovely the lute sounded. It was absolutely thrilling to have achieved something that I'd first dreamed of when I was twenty years old.

With the nicotine patch, it was hard to know what the effect was. Until I finished the whole training process and started writing this book, I never felt anything, really—certainly not the noticeable buzz from drinking a cup of coffee. But I must say that once I got deep into the writing process, it definitely seemed to me—and it might have been a placebo effect—that on days when I wore the patch, I was more focused and productive than on days when I didn't. Maybe it was all in my mind, but I can recall a couple of days in particular when I was really lost and going in circles, only to realize that I'd forgotten that day to put on the patch. Who knows, but for what it's worth, it seemed to help.

As for dear Patsy's boot camp, it never ceased to be hell. I quickly built up enough strength and stamina to almost keep up with most of the others in the class, but even they agreed that boot camp is always hell. I did, however, finally drop below 190 pounds for the first time in five years. And when I ran the Spring Lake Five-Mile Run on May 25, 2013, I cut my previous year's time by more than ten minutes. And my average blood-sugar level, the hemoglobin A1C, also improved significantly.

On Lumosity, my "brain performance index," as they call it, started at 274 on October 12, 2012, and peaked on my last day of training, February 5, 2013, at 1,135—more than four times higher. Compared to others in my age group, my standing also soared, from the 43rd percentile at the beginning to the 93rd percentile by the time I finished. The increase was so dramatic, I had to assume there was some bullcrap factored in just to make me and people like me feel good. Not that they doctored the numbers; I could see every day how my improvement on each task resulted in a slight uptick in my overall score. But just as I was stalling out at a certain level on a particularly tough game, it always seemed that the automatic offering of recommended games would conveniently suggest some new ones, on which I then quickly began climbing again. It also seemed that the program gave a few points each day for simply doing the games, whether or not any improvement had been achieved. And maybe all that makes sense, particularly if they're trying to keep people motivated to keep playing. But the idea that my brain is now performing more than four times better than when I started, or that I'm four times smarter, is of course ridiculous.

Finally, on the dual N-back, my gains were comparable to what Jaeggi and Buschkuehl have seen among people over the age of fifty: I started out struggling, like everybody else, just to master the 2-back level when I began in October, and by January I was routinely hitting 5-back. It was far harder than anything on Lumosity and required much more of the kind of self-motivated determination that is a requirement for a professional writer. Generally speaking, my mood and the amount of sleep I'd had played an obvious role in how well I did, but sometimes, when I felt my worst, tired and grumpy, my scores still improved, and other days, when I felt at my best, they dropped. And by the last few weeks, I definitely stopped improving, and I never did hit 6-back. It kind of reminded me of when I used to play basketball and go bowling as a kid and never got much better. I just sucked at those games, as I did at most sports, which is why I gave up sports in my teen years and only took up jogging, at the age of twenty, because it required absolutely no skill.

But isn't that how life teaches us? We find out what we're good at as children and teenagers, and we stick with it. As soon as we're done with high school, most of us try to get as far away as possible from doing things we suck at. The ones who hated English class forget about reading books, and the ones who found gym class to be torture avoid sports for the rest of their lives. And until I started my training regimen, I stuck with being a journalist, writing a lot and reading a lot, because that was what I was good at. And one of the main things I learned from this training process was that it forced me to do things I sucked at. I found that liberating, and inspiring, and if I improved my cognitive abilities in any way, I believe it was due, at least in part, to simply engaging in learning new things, whether or not they were designed by psychologists specifically to improve intelligence.

So there I was, driving down the New Jersey Turnpike and onto Interstate 95 to the University of Maryland, just outside of Washington, D.C., where Buschkuehl subjected me to another day and a half of torture. And then I waited. Two weeks later, he e-mailed me my results.

The bad news was that on the "surface development" test, where you have to figure out how the flattened pieces of a box would look when assembled, I had actually dropped, from getting 23 percent correct to 13 percent correct. And that compared to the average undergraduate students' performance, based on Buschkuehl and Jaeggi's earlier studies, of 65 percent correct.

So I started out dumber than the kids, and I got even dumber.

But on both the "space relations" and BOMAT tasks, my score was flat, unchanged before and after, at 37 percent and 67 percent, respectively. My score on "space relations" was just a hair below the average student's, while on BOMAT it was just above.

Things went much better on the remaining three tests. On the digit-symbol task, I climbed a bit, from 36 percent to 40 percent. On the "form board" task I climbed a lot, from 40 percent to 55 percent, bringing me dead-even with the average score of the undergraduates. And on the Raven's, the gold standard, I climbed from 67 percent accurate to 78 percent, a single point below the undergraduates.

Step back. Even including my strange collapse on the "surface development" task, my combined score still climbed by 3 percent in absolute numbers. A fairer way to gauge the change, however, is proportionally, since a 3-point gain in absolute numbers would be extraordinary if you started out at 3 and ended up at 6 but would be negligible if you started at 97 and ended up at 100. Proportionally, my combined score on their state-of-the-art set of fluid-intelligence measures climbed by 6 percent. And proportionally, on the gold-standard Raven's, the single best measure of fluid intelligence, my score climbed 16.4 percent.

Does that mean I'm 3 percent smarter, or 6 percent smarter, or 16 percent smarter? Take your pick, but either way, my fluid intelligence measurably improved.

And so what? Those are just numbers on a test. In the end, for all of us, the best test of cognitive abilities is one for which there is no answer key. It's called life.

From 1986 to 1989, Marilyn vos Savant was listed in the *Guinness Book of World Records* as having the highest recorded IQ in the world for women, at 190. Since then, what has she done? She has been writing *Parade* magazine's "Ask Marilyn" advice column. I mean—*really*?

If intelligence is calculated by what we do, you hold in your hands the single best measure of mine. My days of training were filled with purposeful, challenging tasks of all kinds. Boot camp. Lute. N-back. Lumosity. (My editor tells me that some of my e-mails to her at the time were "alarmingly energetic"!) The tasks were hard, but they were fun. I got along better with my wife and daughter. I no longer found myself getting into my car and realizing that I'd forgotten my briefcase. I went on nearly a dozen trips to scientific meetings around the country during the same period, booking all my flights and rental cars and hotels but experiencing none of the stress and sense of being overwhelmed that I'd expected. And then I wrote this book. It sounds pat and clichéd, but what can I tell you?

I feel smarter.

Afterword

Looming over the field of cognitive training in the year since *Smarter* was published in hardcover have been two unanswered questions: Does cognitive training truly increase performance on tests of fluid intelligence, given the negative findings of the meta-analysis (described in chapter 10) by Charles Hulme and Monica Melby-Lervåg? And, even if it does, are there any real-world benefits beyond the ability to take an intelligence test?

Both of those questions have now been answered, so much so that U.S. military and intelligence agencies are investing in new initiatives to enhance the mental fitness of recruits and high-level officers alike.

Perhaps the most extraordinary of the studies showing real-world benefits was published in January 2014, with funding from the National Institute of Aging. Ten years after a group of 2,832 older volunteers participated in a mere ten hours of computerized cognitive training, researchers reported in the *Journal of the American Geriatric Society*, those who were randomly assigned to programs designed to improve their reasoning and reaction times still showed measurable benefits in everyday activities.

That almost sounds implausible: How could ten hours of training still show a positive effect ten years later? But the findings were considered reliable enough for the director of the National Institute of Aging, Richard J. Hodes, to state: "These longer-term results indicate that particular types of cognitive training can provide a lasting benefit a decade later. They suggest that we should continue to pursue cognitive training as an intervention that might help maintain the mental abilities of older people so that they may remain independent and in the community."

Another important new study demonstrating real-world benefits, this time in military personnel, was reported by Andy McKinley, a civilian biomedical engineer who has been studying tDCS at the Air Force Research Laboratory at Wright-Patterson Air Force Base in Ohio. He showed that brief treatments with low-dose electrical stimulation of the brain improved the vigilance and target detection of fighter pilots.

"The military has been looking at how to improve vigilance for the past fifty or sixty years," McKinley told me. "At minimum we get a two-fold improvement in how long a person can maintain optimal performance. We've never seen that with anything else."

A third study demonstrating real-world benefits—in arguably the single most deserving group of people in need of cognitive stimulation: impoverished children in developing countries—was published on April 29, 2014, in the *Proceedings of the National Academy of Sciences*. Argentinean researchers reported that less than six hours of brain games played over the course of ten weeks enabled poor first-graders who attend school irregularly due to family problems to catch up with their regularly attending peers in math and language grades.

"Here's what the critics have asked for," said Jason M. Chein, associate professor of psychology and principal investigator of the Temple University Neurocognition Lab in Philadelphia. "They have said these studies don't translate into real-world benefits. But in the hands of these scientists, the effects look positive."

The study involved 111 impoverished first-graders living in the slums of Buenos Aires. They were taken out of their classrooms for fifteen minutes at a time, up to three times per week, for ten weeks, to play either ordinary computer games or specially designed games intended to increase attention, planning, and working memory. Children who attended school regularly saw no significant gain on their school grades associated with the training. But those whose school attendance was erratic, presumably due to disordered home environments, improved enough in language and math for their grades to catch up with their classmates.

"With very brief training, we improved the language and math grades of children who have problems at home," said the senior author, Andrea P. Goldin, a research scientist at the Integrative Neuroscience Laboratory at the University of Buenos Aires. "That's what we are very excited about, because we are helping to equalize a bit the opportunities these children have."

But what about that meta-analysis by Hulme and Melby-Lervåg? Just about every news article published on the subject of brain training continues to quote that study as "proof" that brain training doesn't really work, that it doesn't stand up to scrutiny. That's why it's so important that four new meta-analyses have now been published, each one concluding that training confers significant benefits.

First, in March 2014, researchers from Northwestern University presented a study combining the results of thirteen previous studies of computerized brain-training in healthy young adults. "We're pretty confident of the main finding that a proper meta-analysis shows a consistent and reliable, if modest, effect of working memory training on fluid intelligence measures," said Paul J. Reber, professor of psychology at Northwestern University.

Two months later, in May, Jaeggi herself presented another new meta-analysis at the annual meeting of the American Psychological

Society. She, Buschkuehl, and others combined the results of twenty previous studies of the N-back in healthy young adults. Even including the results from studies conducted by some of the leading skeptics of training, her meta-analysis showed a significant but modest net benefit on tests of fluid intelligence.

That same month, the journal *Ageing Research Reviews* published another meta-analysis, showing significant benefits of cognitive training in older adults. The following month, in June, the *Journal of Alzheimer's Disease* published a fourth new meta-analysis, showing that training even improved the mental status of seniors suffering from mild cognitive impairment, a transitional stage between normal aging and dementia.

So impressive are findings like these to federal agencies whose job it is to protect the security of the United States that a $12 million program was announced in January 2014 by the Intelligence Advanced Research Projects Activity (IARPA). It will pay for the first year of a planned three-and-a-half-year program called Strengthening Human Adaptive Reasoning and Problem-solving (SHARP).

The SHARP program is studying techniques both ancient and avant-garde, from meditation to tDCS, with an aim toward making intelligence analysts, well, more intelligent. Also on the drawing board are large-scale studies of working memory games.

"If these interventions are actually doing what we think they're doing," said Adam Russell, a neuroscientist and the SHARP program's manager at IARPA, "we should be able to demonstrate that with large numbers of participants, strong metrics and a real-world test battery."

Soldiers are already benefiting from some of the new cognitive treatments. In October 2010 Jessie Kent Fletcher was serving as a Marine scout sniper in Helmand province in Afghanistan when he stepped on an improvised explosive device near the top of a hill. After somersaulting through the air and landing on his back, he saw that both of his lower legs and several fingers were gone. Even more devastating, though, was the loss of his memory and mental focus. "That was the hardest part of the recovery—trying to get back my memory and the functions of day-to-day living," Fletcher told me. "Without your mind, it's really hard to continue moving forward. My attention to detail became so gray. I was just baffled."

After four months of computerized brain training during 2012 at Walter Reed National Military Medical Center's Brain Fitness Center, Fletcher said his cognitive abilities went "from subpar to excellent." Now married to his longtime girlfriend, he lives in Winston-Salem, North Carolina, where they both attend Salem College. During his freshman semester in the fall of 2013, he earned straight A's. "I don't want to live on retirement the rest of my life," Fletcher told me. "I want to be a productive member of society."

The biggest change in my own life since completing work on *Smarter* is that my wife and I are now foster parents to a child who, as of this writing, is seven years old. She was never sent to preschool or kindergarten before she came to us live with us, and couldn't read, write, or draw so much as a happy face. Just ten months later, however, she is now scoring in the 90th percentile on her kindergarten math tests and creating homework assignments for my wife and me to complete.

More than ever, I am convinced that brain training of all kinds has life-changing benefits, and determined to spread the news from scientists that our minds, and our future, should never again be defined by yesterday.

ACKNOWLEDGMENTS

Thanks first to Alberto Costa for convincing me, in December 2009, that the research he and others were doing into finding a treatment to improve the cognitive abilities of people with Down syndrome was a story being ignored by journalists and deserving of a wider audience. I also wish to thank his lovely wife and daughter, Daisy and Tyche, and the families involved in his study, for allowing me into their lives. Thanks also to Fay Ellis, editor of *Neurology Today*, for assigning me that first story about the subject, and for so many other assignments to explore the wonder world that is modern neurologic research. More than three years later, Fay also kindly reviewed the manuscript of this book, as did fellow freelance science reporter Murray Carpenter and my friend Lydia Golub.

At the *New York Times Magazine*, Ilena Silverman assigned and edited my feature article about Costa's work, as well as a later piece about Jaeggi and Buschkuehl's research. The Costa story prompted Jennifer Rudolph Walsh, head of the literary department at William Morris Endeavor, to reach out to me. Jay Mandel, the agent there who took me under his wing, has given me the kind of attention, and got me the kind of deal, writers dream of. Everyone at WME, including Tracy Fisher, Lizzie Thompson, Julianne Wurm, and the inimitable Ari Emanuel, has been awesome.

Caroline Sutton, editor in chief at Hudson Street Press, told me she knew this book needed to be written. Her passion, attention, and wisdom have been an inspiration.

Pamela Weintraub at *Discover* magazine assigned me to write an article about nicotine, from which my account of its uses as a cognitive enhancer, in chapter 5, is partly drawn. She is one of the finest science editors I know.

I pestered a lot of researchers to research this book, and they all deserve my thanks, but especially Susanne Jaeggi and Martin Buschkuehl, who not only agreed to test my fluid intelligence before and after my cognitive training program, but endured visit after visit, interview after interview, by telephone, in person, and via e-mail; and Randy Engle at Georgia Tech, whom I really do like and respect, even though I do not share his deeply negative view of the field. Mike Cole and Todd Braver at Washington University took the time and expense to perform fMRI scans on my brain before and after my training regimen. Felipe Fregni at Harvard agreed to treat me with three sessions of transcranial direct-current stimulation. Linda Gottfredson invited me to attend the International Society of Intelligence Researchers annual meeting. Kate Sullivan, director of the Brain Fitness Center, and Louis French, chief of traumatic brain injury, at Walter Reed National Military Medical Center in Bethesda, Maryland, gave their permission for me to visit and meet their patients. And Harold Hawkins at the Office of Naval Research enabled me to attend the extraordinary meeting of his grantees described in chapter 10. I also wish to thank all those researchers whose names do not appear in this book, but who nevertheless spoke with me, sent me their studies, and otherwise helped me get a grip on this enormously difficult subject.

I regret not being able to find a place in the book to include the stories of the four wounded warriors who met with me at Walter Reed to share their moving accounts of how they have sought to overcome the effects of their traumatic brain injuries. To Jessie Kent Fletcher, Corporal E4 with the U.S. Marine Corps, who lost his legs but not his courage in Afghanistan; Master Sergeant Jill Westeyn, who played oboe in the U.S. Air Force Band until a loose hook hanging from the ceiling of a military transport plane fell onto her head, and who regained her cognitive abilities through training but not her original sense of hearing; Octavio Tapia, U.S. Army, private first class, who has worked heroically to overcome the effects of an arteriovenous malformation in his brain; and Cody Banks, who was serving as a squad leader in the U.S. Marine Corps on July 3, 2011, when an improvised explosive device went off in the town of Salaam Bazaar, Afghanistan, causing him a traumatic brain injury from which he is still fighting to recover: I thank you for your service to our country, and for putting up with my prying questions.

I also wish to acknowledge the kindness of Mimi Chesslin and the late Charles Feigelson, the parents of my childhood friend, Dan. Back in the 1960s, when I was growing up, they always made me feel welcome in their home, where the marble chess board on their living-room table, the Modigliani print hanging on the wall, the Beethoven playing on the phonograph, and the books lining their shelves had an immeasurable influence on my transformation from a "slow learner" into a professional writer.

Much of this book was written either in the majestic Rose Main Reading Room of the New York Public Library's main branch or in its inner sanctum, room 300, holding the Wallach division of art, prints, and photographs. I thank the librarians and guards there for expanding my own mind's workspace.

Finally, as always, to my family in Maine and Florida; to my friends in and beyond New Jersey; to my beautiful, talented, and gracious wife, Alice, who has shared the thrills and despairs of my life as a freelance journalist for over two decades now; and to our lovely daughter, Annie, whose strength and brilliance dazzle me: all my love and thanks.

NOTES

Introduction

xiv **Bright Kids NYC now has as many as five hundred children enrolled:** http://www.brightkidsnyc.com/2013/01/olsat-bootcamp-waitlist-now-available/. Accessed on July 13, 2013.

xv **more kids overall, and more rich ones than ever, were accepted:** Anna M. Phillips, "After Number of Gifted Soars, a Fight for Kindergarten Slots," *New York Times*, April 13, 2012. See also Elissa Gootman, "More Children Take the Tests for Gifted Programs, and More Qualify," *New York Times*, May 5, 2009.

xv **even more kids overall, and more rich kids in particular, passed the test:** Al Baker, "More in New York City Qualify as Gifted After Error Is Fixed," *New York Times*, April 19, 2013.

xv **94 percent of the children who prepped with Bright Kids scored in the 90th percentile:** http://www.brightkidsnyc.com/2013/04/bright-kids-nyc-delivers-record-results-on-the-gifted-and-talented-exam/. Accessed on July 13, 2013.

xvi **it was in the United States, after all, that the pseudoscience of eugenics had its birthplace:** See the extraordinary and shocking book by investigative journalist Edwin Black, *War Against the Weak: Eugenics and America's Campaign to Create a Master Race* (Washington, D.C.: Dialog Press, 2003).

xvi **Four years later, the writer of that dissertation, Jason Richwine, authored a study for the Heritage Foundation:** Ashley Parker and

Julia Preston, "Paper on Immigrant I.Q. Dogs Critic of Overhaul," *New York Times*, May 8, 2013.

xvii **"rein in emotional impulse":** Daniel Goleman, *Emotional Intelligence* (New York: Bantam Books, 1995).

xvii **ten thousand hours:** Malcolm Gladwell, *Outliers: The Story of Success* (New York: Little, Brown, 2008).

xviii **characteristics like self-control:** Paul Tough, *How Children Succeed: Grit, Curiosity, and the Hidden Power of Character* (New York: Houghton Mifflin Harcourt, 2012). See also Duckworth AL, Peterson C, et al., "Grit: Perseverance and passion for long-term goals." *J Pers Soc Psychol.* 2007 Jun; 92 (6):1087–1101.

xviii **Center for Talented Youth at Johns Hopkins:** http://en.wikipedia.org/wiki/Center_for_Talented_Youth.

xviii **A recent study of 1,116,442 Swedish men:** Batty GD, Gale CR, et al., "IQ in early adulthood, socioeconomic position, and unintentional injury mortality by middle age: A cohort study of more than 1 million Swedish men." *Am J Epidemiol.* 2009 Mar 1; 169 (5): 606–15.

xix **study, of Scottish adults born in 1921:** Hart CL, Taylor MD, et al., "Childhood IQ, social class, deprivation, and their relationships with mortality and morbidity risk in later life." *Psychosomatic Medicine.* 2003 Sep–Oct; 65 (5): 877–83.

xix **murdered, developing high blood pressure, having a stroke or heart attack:** Battya GD, Dearyb IJ, "The association between IQ in adolescence and a range of health outcomes at 40 in the 1979 US National Longitudinal Study of Youth." *Intelligence.* 2009 Nov–Dec; 37 (6): 573–80.

xix **even to early menopause:** Kuh D, Butterworth S, et al., "Childhood cognitive ability and age at menopause: Evidence from two cohort studies." *Menopause.* 2005 Jul–Aug; 12 (4): 475–82.

xix **800,000 children and adults in the United States receiving Social Security income due to a diagnosed intellectual disability:** National Council on Disabilities, http://www.ncd.gov/publications/2013/20130315/20130315_Ch2.

xix **250,000 service members diagnosed with a traumatic brain injury since 2000:** James Dao, "Rules Eased for Veterans' Brain Injury Benefits," *New York Times*, December 7, 2012.

xix **5 million Americans . . . Alzheimer's:** Alzheimer's Association, http://www.alz.org/alzheimers_disease_facts_and_figures.asp.

xix **Alzheimer's . . . "cognitive reserve":** Yeo RA, Arden R, et al., "Alzheimer's disease and intelligence." *Current Alzheimer Research.* 2011 Jun; 8 (4): 345–53.

xx **major depression or schizophrenia:** Vinogradov S, Fisher M, et al., "Cognitive training for impaired neural systems in neuropsychiatric illness." *Neuropsychopharmacology.* 2012 Jan; 37 (1): 43–76.

xx **first new answer in a century:** Jaeggi SM, Buschkuehl M, "Improving fluid intelligence with training on working memory." *PNAS.* 2008 May 13; 105 (19): 6829–33.

xxi **the headline of an editorial:** Sternberg R, "Increasing fluid intelligence is possible after all." *PNAS.* 2008 May 13; 105 (19): 6791–92.

xxii **four . . . studies . . . finding no benefit of cognitive training:** Redick TS, Shipstead Z, et al., "No evidence of intelligence improvement after working memory training: A randomized, placebo-controlled study." *J Exp Psychol Gen.* 2013 May; 142 (2): 359–79. Smith SP, Stibric M, et al., "Exploring the effectiveness of commercial and custom-built games for cognitive training." *Computers in Human Behavior.* 2013 Nov; 29 (6): 2388–93. Thompson TW, Waskom ML, "Failure of working memory training to enhance cognition or intelligence." *PLoS One.* 2013 May 22; 8 (5): e63614. Chooi W-T, Thompson LA, "Working memory training does not improve intelligence in healthy young adults." *Intelligence.* 2012 Nov–Dec; 40 (6): 531–42.

xxii **seventy-five other randomized, placebo-controlled studies:** Academic skeptics and advocates of cognitive training passionately dispute which studies show positive effects and which show negative effects, as will be described in detail in chapters 8 and 10. Some experts cite more studies as showing positive evidence, and some believe that far fewer hold up to scrutiny. Most of this book is devoted to describing and assessing those studies. I include among the seventy-five studies showing positive effects only those with randomized, placebo-controlled designs published in peer-reviewed scientific journals. Included are studies not only of the dual N-back task, as originally studied by Jaeggi and colleagues in their 2008 paper, but also other kinds of working-memory training, as well as other cognitive training programs, such as those designed by Michael Merzenich of Posit Science. In addition, I include among the seventy-five studies those that used mindfulness meditation, music training, and physical exercise. While only twenty-two of the positive studies found benefits for fluid intelligence per se, the remaining fifty-three studies all found other significant benefits on measures of attention, working memory, reading, or other important intellectual abilities. What all of these studies have in common is that they were relatively brief and straightforward and simple to implement, required no more than two to four hours of training per week, and reported statistically significant cognitive benefits. Here I present the fifty-three studies that showed benefits that did not include increased fluid intelligence; in the next note, I present the twenty-two additional studies that found benefits for fluid intelligence or reasoning.

Alloway TP, Alloway RG, "The efficacy of working memory training in improving crystallized intelligence." *Nature Precedings*. 2009 Sep. http://precedings.nature.com/documents/3697/version/1.

Anderson S, White-Schwoch T, et al., "Reversal of age-related neural timing delays with training." *PNAS*. 2013 Mar 12; 110 (11): 4357–62.

Angevaren M, Aufdemkampe G, et al., "Physical activity and enhanced fitness to improve cognitive function in older people without known cognitive impairment." *Cochrane Database Syst Rev*. 2008 (3): CD005381.

Ball K, Berch DB, et al., "Effects of cognitive training interventions with older adults: A randomized controlled trial." *JAMA*. 2002 Nov 13; 288 (18): 2271–81.

Bell M, Bryson G, et al., "Cognitive remediation of working memory deficits: Durability of training effects in severely impaired and less severely impaired schizophrenia." *Acta Psychiatr Scand*. 2003; 108: 101–9.

Bennett SJ, Holmes J, et al., "Computerized memory training leads to sustained improvement in visuospatial short-term memory skills in children with Down syndrome." *Am J Intel Dev Disab*. 2013; 118 (3): 179–92.

Boron JB, Willis SL, et al., "Cognitive training gains as a predictor of mental status." *J Gerontol*. 2007 Jan; 62B (1): P45–51.

Brehmer Y, Westerberg H, et al., "Working-memory training in younger and older adults: Training gains, transfer, and maintenance." *Front Hum Neurosci.* 2012 Mar; 6: 63.

Buschkuehl M, Jaeggi SM, et al., "Impact of working memory training on memory performance in old-old adults." *Psychol Aging.* 2008 Dec; 23 (4): 745–53.

Chein JM, Morrison AB, "Expanding the mind's workspace: Training and transfer effects with a complex working memory span task." *Psychon Bull Rev.* 2010 Apr; 17 (2): 193–99.

Dahlin KIE, "Effects of working memory training on reading in children with special needs." *Reading and Writing.* 2011; 24: 479–91.

Erickson KI, Voss MW, et al., "Exercise training increases size of hippocampus and improves memory." *PNAS.* 2011 Feb 15; 108 (7): 3017–22.

Fisher M, Holland C, et al., "Using neuroplasticity-based auditory training to improve verbal memory in schizophrenia." *Am J Psychiatry.* 2009 Jul; 166 (7): 805–11.

Gray SA, Chaban P, et al., "Effects of a computerized working memory training program on working memory, attention, and academics in adolescents with severe LD and comorbid ADHD: A randomized controlled trial." *J Child Psychol Psychiatry.* 2012 Dec; 53 (12): 1277–84.

Green CT, Long DL, et al., "Will working memory training generalize to improve off-task behavior in children with attention-deficit/hyperactivity disorder?" *Neurotherapeutics.* 2012 Jul; 9 (3): 639–48.

Hardy KK, Willard VW, et al., "Working memory training in survivors of pediatric cancer: A randomized pilot study." *Psycho-Oncology.* 2012 Dec 2 [Epub ahead of print].

Hawkins HL, Kramer AF, et al., "Aging, exercise, and attention." *Psychol Aging.* 1992 Dec; 7 (4): 643–53.

Heinzel S, Schulte S, et al., "Working memory training improvements and gains in non-trained cognitive tasks in young and older adults." *Neuropsychol Dev Cogn B Aging Neuropsychol Cogn.* 2013 May 2 [Epub ahead of print].

Holmes J, Gathercole SE, et al., "Working memory deficits can be overcome: Impacts of training and medication on working memory in children with ADHD." *Appl Cognit Psychol.* 2010 Sep; 24 (6): 827–36.

Holmes J, Gathercole SE, et al., "Adaptive training leads to sustained enhancement of poor working memory in children." *Dev Sci*. 2009 Jul; 12 (4): F9–15.

Houben K, Wiers RW, et al., "Getting a grip on drinking behavior: Training working memory to reduce alcohol abuse." *Psychol Sci*. 2011 Jul; 22 (7): 968–75.

Kesler S, Hadi Hosseini SM, et al., "Cognitive training for improving executive function in chemotherapy-treated breast cancer survivors." *Clin Breast Cancer*. 2013 Aug; 13 (4): 299–306.

Klingberg T, Fernell E, et al., "Computerized training of working memory in children with ADHD—a randomized, controlled trial." *J Am Acad Child Adolesc Psychiatry*. 2005; 44 (2): 177–86.

Klusmann V, Evers A, et al., "Complex mental and physical activity in older women and cognitive performance: A 6-month randomized controlled trial." *J Gerontol A Biol Sci Med Sci*. 2010 Jun; 65 (6): 680–88.

Kramer AF, Hahn S, et al., "Ageing, fitness and neurocognitive function." *Nature*. 1999 Jul 29; 400 (6743): 418–19.

Kray J, Karbach J, "Can task-switching training enhance executive control functioning in children with attention deficit/-hyperactivity disorder?" *Front Hum Neurosci*. 2011; 5: 180.

Kundu B, Sutterer DW, et al., "Strengthened effective connectivity underlies transfer of working memory training to tests of short-term memory and attention." *J Neurosci*. 15 May 2013; 33 (20): 8705–15.

Li SC, Schmiedek F, et al., "Working memory plasticity in old age: Practice gain, transfer, and maintenance." *Psychol Aging*. 2008 Dec; 23 (4): 731–42.

Lilienthal L, Tamez E, et al., "Dual n-back training increases the capacity of the focus of attention." *Psychon Bull Rev*. 2013 Feb; 20 (1): 135–41.

Liu-Ambrose T, Nagamatsu LS, et al., "Resistance training and executive functions: A 12-month randomized controlled trial." *Arch Intern Med*. 2010; 170 (2): 170–78.

Loosli SV, Buschkuehl M, et al., "Working memory training improves reading processes in typically developing children." *Child Neuropsychol*. 2012; 18 (1): 62–78.

McGurk SR, Mueser KT, et al., "Cognitive training and supported employment for persons with severe mental illness: One-year results from a randomized controlled trial." *Schiz Bull.* 2005; 31 (4): 898–909.

Minear M, Shah P, "Training and transfer effects in task switching." *Mem Cognit.* 2008 Dec; 36 (8): 1470–83.

Nagamatsu LS, Handy TC, et al., "Resistance training promotes cognitive and functional brain plasticity in seniors with probable mild cognitive impairment." *Arch Intern Med.* 2012 Apr 23; 172 (8): 666–68.

Nagamatsu LS, Chan A, et al., "Physical activity improves verbal and spatial memory in older adults with probable mild cognitive impairment: A 6-month randomized controlled trial." *J Aging Res.* 2013; 2013: article ID 861893.

Olesen PJ, Westerberg H, Klingberg T. "Increased prefrontal and parietal activity after training of working memory." *Nat Neurosci.* 2004 Jan; 7 (1): 75–79.

Owens M, Koster EH, et al., "Improving attention control in dysphoria through cognitive training: Transfer effects on working memory capacity and filtering efficiency." *Psychophysiology.* 2013 Mar; 50 (3): 297–307.

Prins PJ, Dovis S, et al., "Does computerized working memory training with game elements enhance motivation and training efficacy in children with ADHD?" *Cyberpsychol Behav Soc Netw.* 2011 Mar; 14 (3): 115–22.

Reed JA, Maslow AL, et al., "Examining the impact of 45 minutes of daily physical exercise on cognitive ability, fitness performance, and body composition of African American youth." *J Phys Act Health.* 2013 Feb; 10 (2): 185–97.

Richmond LL, Morrison AB, et al., "Working memory training and transfer in older adults." *Psychol Aging.* 2011 Dec; 26 (4): 813–22.

Salminen T, Strobach T, et al., "On the impacts of working memory training on executive functioning." *Front Hum Neurosci.* 2012; 6: 166.

Schweizer S, Hampshire A, et al., "Extending brain-training to the affective domain: Increasing cognitive and affective executive control through emotional working memory training." *PLoS One.* 2011; 6 (9): e24372.

Schweizer S, Grahn J, et al., "Training the emotional brain: Improving affective control through emotional working memory training." *J Neurosci.* 2013 Mar 20; 33 (12): 5301–11.

Smith GE, Housen P, et al., "A cognitive training program based on principles of brain plasticity: Results from the Improvement in Memory with Plasticity-based Adaptive Cognitive Training (IMPACT) study." *J Am Geriatr Soc.* 2009 Apr; 57 (4): 594–603.

Subramaniam K, Luks TL, et al., "Computerized cognitive training restores neural activity within the reality monitoring network in schizophrenia." *Neuron.* 2012 Feb 23; 73: 842–53.

Tang Y-Y, Ma Y, et al., "Short-term meditation training improves attention and self-regulation." *PNAS.* 2007 Oct 23; 104 (43): 17152–56.

Tang Y-Y, Lu Q, et al., "Mechanisms of white matter changes inducted by meditation." *PNAS.* 2012 Jun 26; 109 (26): 10570–74.

Thorell LB, Lindqvist S, et al., "Training and transfer effects of executive functions in preschool children." *Dev Sci.* 2009 Jan; 12 (1): 106–13.

Van der Molen MJ, Van Luit JE, et al., "Effectiveness of a computerised working memory training in adolescents with mild to borderline intellectual disabilities." *J Intellect Disabil Res.* 2010 May; 54 (5): 433–47.

Vartaniana O, Jobidona ME, "Working memory training is associated with lower prefrontal cortex activation in a divergent thinking task." *Neuroscience.* 2013 Apr 16; 236: 186–94.

von Bastian CC, Langer N, et al., "Effects of working memory training in young and old adults." *Mem Cognit.* 2013 May; 41 (4): 611–24.

Wolinsky FD, Vander Weg MW, et al., "A randomized controlled trial of cognitive training using a visual speed of processing intervention in middle aged and older adults." *PLoS One.* 2013 May 1; 8 (5): e61624.

Zinke K, Zeintl M, "Working memory training and transfer in older adults: Effects of age, baseline performance, and training gains." *Dev Psychol.* 2013 May 20 [Epub ahead of print].

xxii **Twenty-two of those studies specifically found improvements in fluid intelligence or reasoning:** Here follows a list of those twenty-two studies.

Basak C, Boot WR, et al., "Can training in real-time strategy video game attenuate cognitive decline in older adults?" *Psychol Aging.* 2008; 23 (4): 765–77.

Bergman NS, Soderqvist S, et al., "Gains in fluid intelligence after training non-verbal reasoning in 4-year-old children: A controlled, randomized study." *Dev Sci.* 2011 May; 14 (3): 591–601.

Borella E, Carretti B, et al., "Working memory training in older adults: Evidence of transfer and maintenance effects." *Psychol Aging.* 2010; 25 (4): 767–778.

Carretti B, Borella E, et al., "Gains in language comprehension relating to working memory training in healthy older adults." *Int J Geriatr Psychiatry.* 2013 May; 28 (5): 539–46.

Herrnstein RJ, Nickerson RS, et al., "Teaching thinking skills." *Am Psychol.* 1986 Nov; 41 (11): 1279–89.

Jaeggi SM, Studer-Luethi B, et al., "The relationship between n-back performance and matrix reasoning—implications for training and transfer." *Intelligence.* 2010; 38: 625–35.

Jaeggi SM, Buschkuehl M, et al., "Short- and long-term benefits of cognitive training." *PNAS.* 2011 Jun 21; 108 (25): 10081–86.

Jausovec N, Jausovec K, "Working memory training: Improving intelligence—changing brain activity." *Brain Cogn.* 2012; 79: 96–106.

Karbach J, Kray J, "How useful is executive control training? Age differences in near and far transfer of task-switching training." *Dev Sci.* 2009 Nov; 12 (6): 978–90.

Mackey AP, Hill SS, et al., "Differential effects of reasoning and speed training in children." *Dev Sci.* 2011 May; 14 (3): 582–90.

Moreno S, Bialystok E, et al., "Short-term music training enhances verbal intelligence and executive function." *Psychol Sci.* 2011 Nov; 22 (11): 1425–33.

Neville HJ, Stevens C, et al., "Family-based training program improves brain function, cognition, and behavior in lower socioeconomic status preschoolers." *PNAS.* 2013 Jul 1 [Epub ahead of print].

Roughan L, Hadwin JA, "The impact of working memory training in young people with social, emotional and behavioral difficulties." *Learning and Individual Differences.* 2011 Dec; 21 (6): 759–64.

Rudebeck SR, Bor D, et al., "A potential spatial working memory training task to improve both episodic memory and fluid intelligence." *PLoS One.* 2012; 7 (11): e50431.

Rueda MR, Checa P, et al., "Enhanced efficiency of the executive attention network after training in preschool children: Immediate and after two months effects." *Dev Cogn Neurosci.* 2012 Feb 15; 2 Suppl 1: S192–204.

Rueda MR, Rothbart MK, et al., "Training, maturation, and genetic influences on the development of executive attention." *PNAS*. 2005 Oct 11; 102 (41) 14931–36.

Schmiedek F, Lövdén M, et al., "Hundred days of cognitive training enhance broad cognitive abilities in adulthood: Findings from the COGITO study." *Front Aging Neurosci*. 2010 Jul 13; 2.

Soderqvist S, Nutley SB, et al., "Computerized training of non-verbal reasoning and working memory in children with intellectual disability." *Front Hum Neurosci*. 2012; 6: 271.

Stephenson CL, Halpern DF, "Improved matrix reasoning is limited to training on tasks with a visuospatial component." *Intelligence*. 2013; 41: 341–57.

Von Bastian CC, Oberauer K, "Distinct transfer effects of training different facets of working memory capacity." *J Mem Lang*. 2013 Jul; 69 (1): 36–58.

Wolf D, Fischer FU, et al., "Structural integrity of the corpus callosum predicts long-term transfer of fluid intelligence-related training gains in normal aging." *Hum Brain Mapp*. 2012 Sep 11. [Epub ahead of print].

Zhao X, Wang YX, et al., "Effect of updating training on fluid intelligence in children." *Chin Sci Bull*. 2011 Jul; 56 (21): 2202–5.

xxii **Parkinson's disease:** París AP, Saleta HG, et al., "Blind randomized controlled study of the efficacy of cognitive training in Parkinson's disease." *Mov Disord*. 2011 Jun; 26 (7): 1251–58.

xxii **about emotional intelligence, short-term cognitive training has been shown to pay off:** Schweizer S, Grahn J, et al., "Training the emotional brain: Improving affective control through emotional working memory training." *J Neurosci*. 2013 Mar 20; 33 (12): 5301–11.

Chapter 1: Expanding the Mind's Workspace

1 **where in the brain problems requiring working memory are solved:** Klingberg T, Kawashima R, et al., "Activation of multi-modal cortical areas underlies short-term memory." *Eur J Neurosci*. 1996 Sep; 8 (9): 1965–71.

2 **Most famously, K. Anders Ericsson and colleagues:** Ericsson KA, Chase WG, et al., "Acquisition of a memory skill." *Science*. 1980 Jun 6; 208: 1181–82.

3 **classic 1956 paper by psychologist George A. Miller:** Miller GA, "The magical number seven, plus or minus two: Some limits on our capacity for processing information." *Psychol Rev.* 1956 Mar; 63 (2): 81–97.

3 **These kinds of memory tricks:** Joshua Foer, *Moonwalking with Einstein: The Art and Science of Remembering Everything* (New York: Penguin Press, 2011).

4 **As one researcher has called it:** Chein JM, Morrison AB, "Expanding the mind's workspace: Training and transfer effects with a complex working memory span task." *Psychon Bull Rev.* 2010 Apr; 17 (2): 193–99.

5 **In the early 1980s, . . . Merzenich published studies:** See, for instance, Merzenich MM, Nelson RJ, et al., "Somatosensory cortical map changes following digit amputation in adult monkeys." *J Comp Neuro.* 1984 Apr 20; 224 (4): 591–605; and Merzenich MM, Kaas JH, et al., "Topographic reorganization of somatosensory cortical areas 3b and 1 in adult monkeys following restricted deafferentation." *Neuroscience.* 1983 Jan, 8 (1): 33–55.

5 **Dyslexic children, he found, could be trained:** Temple E, Deutsch GK, et al., "Neural deficits in children with dyslexia ameliorated by behavioral remediation: Evidence from functional MRI." *PNAS* 2003 Mar 4; 100 (5): 2860–65.

5 **drivers in their seventies could likewise be trained:** Wolinsky FD, Vander Weg MW, et al., op.cit.

6 **Klingberg enrolled fourteen children . . . diagnosed with ADHD:** Klingberg T, Forssberg H, et al., "Training of working memory in children with ADHD." *J Clin Experi Neuropsych.* 2002; 24 (6): 781–91.

7 **the two are not synonymous:** Torkel Klingberg, *The Learning Brain: Memory and Brain Development in Children* (New York: Oxford University Press, 2013).

12 **They finally published the results of their study:** Jaeggi SM, Buschkuehl M, et al., "Improving fluid intelligence with training on working memory." *PNAS.* 2008 May 13; 105 (19): 6829–33.

12 **headlines in newspapers:** Roger Highfield, " 'Brain Training' Games Do Work, Study Finds," *Telegraph*, April 28, 2008. Nicholas Bakalar, "Memory Training Shown to Turn Up Brainpower," *New York Times*, April 29, 2008.

12 **accompanying celebratory commentary:** Sternberg R, "Increasing fluid intelligence is possible after all." *PNAS*. 2008 May 13; 105 (19): 6791–92.

12 **an article about drugs being tested to increase intelligence in people with Down syndrome:** Dan Hurley, "A Drug for Down Syndrome," *New York Times Magazine*, July 29, 2011.

13 **hundreds of subsequent studies citing it:** A search for the study on Google shows that 539 studies have cited it as of July 14, 2013.

13 **Chein had seen improvements in cognitive abilities:** Chein JM, Morrison AB, "Expanding the mind's workspace: Training and transfer effects with a complex working memory span task." *Psychon Bull Rev*. 2010 Apr; 17 (2): 193–99.

18 **Glenn Schellenberg . . . two papers:** See Moreno S, Bialystok E, op. cit.; and Schellenberg EG, "Music lessons enhance IQ." *Psychol Sci*. 2004 Aug; 15 (8): 511–14.

Chapter 2: Measure of a Man

30 **women, whose total brain size . . . averages about 10 percent smaller than men's:** Ho KC, Roessmann U, et al., "Analysis of brain weight." *Arch Pathol Lab Med*. 1980; 104 (12): 640–45.

30 **Women actually tend to have more gray matter:** Cosgrove KP, Mazure CM, et al., "Evolving knowledge of sex differences in brain structure, function, and chemistry." *Biol Psychiatry*. 2007 Oct 15; 62 (8): 847–55.

30 **Men . . . perform better on visuospatial tasks, . . . women . . . in verbal fluency:** Wai J, Cacchio M, et al., "Sex differences in the right tail of cognitive abilities: A 30 year examination." *Intelligence*. 2010; 38: 412–23.

30 **the proportion of boys scoring above 700 on the math portion of the SAT:** Benbow CP, Stanley JC, "Sex differences in mathematical reasoning ability: More facts." *Science*. 1983 Dec 2; 222 (4627): 1029–31.

30 **that ratio plummeted to less than four times higher by 2010:** Wai J, Cacchio M, op. cit.

30 **males are slightly more likely than females to be cognitively disabled:** Maulik PK, Harbour CK, "Epidemiology of Intellectual Disability." 2013. In JH Stone, M Blouin, editors, *International*

Encyclopedia of Rehabilitation. http://cirrie.buffalo.edu/encyclopedia/en/article/144/.

30 **a study by researchers at Washington University in St. Louis:** Cole MW, Yarkoni T, et al., "Global connectivity of prefrontal cortex predicts cognitive control and intelligence." *J Neurosci*. 2012 Jun 27; 32 (26): 8988–99.

Chapter 3: A Good Brain Trainer Is Hard to Find

37 **Dr. Kawashima's Brain Training:** See https://en.wikipedia.org/wiki/Brain_Age:_Train_Your_Brain_in_Minutes_a_Day!

37 **Only a few studies, in fact, have ever found such benefits:** See, for example, Nouchi R, Taki Y, et al., "Brain training game boosts executive functions, working memory and processing speed in the young adults: A randomized controlled trial." *PLoS One*. 2013; 8 (2): e55518. Brem MH, Lehrl S, et al., "Stop of loss of cognitive performance during rehabilitation after total hip arthroplasty-prospective controlled study." *J Rehabil Res Dev*. 2010; 47 (9): 891–98.

38 **By 2003 Cogmed already had its first paying customers in Sweden:** The chronology in this paragraph is described at http://www.cogmed.com/history.

43 **in the journal *Neurotherapeutics*, Schweitzer's study:** Green CT, Long DL, op. cit.

44 **scores had dropped by an average of 10.3 points:** Krull KR, Zhang N, et al., "Long-term decline in intelligence among adult survivors of childhood acute lymphoblastic leukemia treated with cranial radiation." *Blood*. 2013 Jul 25; 122 (4): 550–53.

45 **a pilot study comparing Cogmed to a placebo:** Hardy KK, Willard VW, op. cit.

48 **By April 2013, they claimed 40 million members:** http://en.wikipedia.org/wiki/Lumosity.

49 **Lumosity has already been the subject of fifteen studies:** See http://hcp.lumosity.com/research/completed. Accessed on August 13, 2013.

49 **continues to be used in dozens of ongoing studies:** See http://hcp.lumosity.com/research. Accessed on August 13, 2013.

49 **Kesler . . . published two pilot studies of Lumosity:** Kesler SR,

Sheau K, et al., "Changes in frontal-parietal activation and math skills performance following adaptive number sense training: Preliminary results from a pilot study." *Neuropsychol Rehabil.* 2011 Aug; 21 (4): 433–54. Kelser SR, Lacayo NJ, et al., "A pilot study of an online cognitive rehabilitation program for executive function skills in children with cancer-related brain injury." *Brain Inj.* 2011; 25 (1): 101–12.

49 **Kesler published her latest study of Lumosity:** Kesler S, Hadi Hosseini SM, op. cit.

51 **the firm released its first public analysis of that database:** Sternberg, DA, Ballard K, et al., "The largest human cognitive performance dataset reveals insights into the effects of lifestyle factors and aging." *Front Hum Neurosci.* 2013 June 20; 7: 292.

53 **Merzenich played a key role in the development of cochlear implants:** See http://en.wikipedia.org/wiki/Michael_Merzenich and http://www.positscience.com/why-brainhq/world-class-science/dr-michael-merzenich.

54 **Two recent studies led by Sophia Vinogradov:** Fisher M, Holland C, op. cit., and Subramaniam K, Luks TL, op. cit.

54 **the National Institute of Mental Health convened a meeting:** "Cognitive Training in Mental Disorders: Advancing the Science." A summary of the meeting can be found at http://www.nimh.nih.gov/research-priorities/scientific-meetings/2012/cognitive-training-in-mental-disorders-advancing-the-science/index.shtml.

56 **Improvement in Memory with Plasticity-based Adaptive Cognitive Training (IMPACT) study:** Smith GE, Housen P, op. cit.

56 **A second study . . . 681 people in two age groups:** Wolinsky FD, Vander Weg MW, op. cit.

57 **self-rated health and other measures that persisted for five years:** Wolinsky FD, Mahncke H, et al., "Speed of processing training protects self-rated health in older adults: Enduring effects observed in the multi-site ACTIVE randomized controlled trial." *Int Psychogeriatr.* 2010 May; 22 (3): 470–78.

57 **none of the cognitive training methods . . . appeared to lower people's risk for [dementia]:** Unverzaqt FW, Guey LT, et al., "ACTIVE cognitive training and rates of incident dementia." *J Int Neuropsychol Soc.* 2012 Jul; 18 (4): 669–77.

61 **trainers for LearningRx have also posted complaints:** Such complaints can be easily found by googling the terms "LearningRx" and "complaints."

63 **Oliver W. Hill Jr. . . . recently completed a $1 million study:** Hill O, Serpell Z, et al., "Improving Minority Student Mathematics Performance through Cognitive Training." In LA Flowers, J Moore, LO Flowers, editors, *The Evolution of Learning: Science, Technology, Engineering, and Mathematics Education at Historically Black Colleges and Universities* (Lanham, Md.: University Press of America, in press).

66 **they reported their extraordinary results:** Green CS, Bavelier D, "Action video game modifies visual selective attention." *Nature*. 2003 May 29; 423 (6939): 534–37.

67 **similar effects on auditory attention:** Green CS, Pouget A, Bavelier D, "Improved probabilistic inference as a general learning mechanism with action video games." *Curr Biol*. 2010 Sep 14; 20 (17): 1573–79.

67 **perceive subtle differences in shades of gray:** Li R, Plat U, et al., "Enhancing the contrast sensitivity function through action video game training." *Nat Neurosci*. 2009 May; 12 (5): 549–51.

67 **poor contrast sensitivity . . . among the strongest risk factors:** Swedenll WR, Ensrud KE, et al., "Indicators of 'healthy aging' in older women (65–69 years of age). A data-mining approach based on prediction of long-term survival." *BMC Geriatr*. 2010 Aug 17; 10: 55.

Chapter 4: Old-School Brain Training

70 **PubMed:** http://www.ncbi.nlm.nih.gov/pubmed.

70 **Blueberries . . . studies in elderly rats:** Krikorian R, Shidler MD, et al., "Blueberry supplementation improves memory in older adults." *J Agric Food Chem*. 2010 Apr 14; 58 (7): 3996–4000.

70 **2012 review by researchers at Hebrew University:** Nachum-Biala Y, Troen AM, "B-vitamins for neuroprotection: Narrowing the evidence gap." *Biofactors*. 2012 Mar–Apr; 38 (2): 145–50.

71 ***Cochrane Database of Systematic Reviews*:** Malouf R, Grimley Evans J, "Folic acid with or without vitamin B12 for the prevention and treatment of healthy elderly and demented people." *Cochrane Database Syst Rev*. 2008 Oct 8; (4): CD004514.

71 **creatine . . . four randomized, blinded trials:** Benton D, Donohoe

R, “The influence of creatine supplementation on the cognitive functioning of vegetarians and omnivores.” *Br J Nutr.* 2011 Apr; 105 (7): 1100–1105. Rawson ES, Lieberman HR, et al., “Creatine supplementation does not improve cognitive function in young adults.” *Physiol Behav.* 2008 Sep 3; 95 (1–2): 130–34. Rae C, Digney AL, et al., “Oral creatine monohydrate supplementation improves brain performance: A double-blind, placebo-controlled, cross-over trial.” *Proc Biol Sci.* 2003 Oct 22; 270 (1529): 2147–50. McMorris T, Mielcarz G, et al., “Creatine supplementation and cognitive performance in elderly individuals.” *Neuropsychol Dev Cogn B Aging Neuropsychol Cogn.* 2007 Sep; 14 (5): 517–28.

71 **Cochrane review of omega-3 PUFA supplements:** Sydenham E, Dangour AD, et al., “Omega 3 fatty acid for the prevention of cognitive decline and dementia.” *Cochrane Database Syst Rev.* 2012 June 13; 6: CD005379.

72 **fish oil supplements for pregnant women. . . . A 2003 study:** Helland IB, Smith L, et al., “Maternal supplementation with very-long-chain n-3 fatty acids during pregnancy and lactation augments children’s IQ at 4 years of age.” *Pediatrics.* 2003 Jan; 111 (1): e39–44.

72 **in 2008, the same group published a follow-up paper:** Helland IB, Smith L, et al., “Effect of supplementing pregnant and lactating mothers with n-3 very-long-chain fatty acids on children’s IQ and body mass index at 7 years of age.” *Pediatrics.* 2008 Aug; 122 (2): e472–79.

72 **study, this one by researchers at the University of Copenhagen:** Asserhøj M, Nehammer S, et al., “Maternal fish oil supplementation during lactation may adversely affect long-term blood pressure, energy intake, and physical activity of 7-year-old boys.” *J Nutri.* 2009 Feb; 139 (2): 293–304.

73 **The same group reported in 2011:** Cheatham CL, Nerhammer AS, et al., “Fish oil supplementation during lactation: Effects on cognition and behavior at 7 years of age.” *Lipids.* 2011 Jul; 46 (7): 637–45.

73 **The Mediterranean diet:** Martínez-Lapiscina EH, Clavero P, et al., “Mediterranean diet improves cognition: The PREDIMED-NAVARRA randomized trial.” *J Neurol Neurosurg Psychiatry.* 2013 May 13 [Epub ahead of print].

74 **mothers who choose to breastfeed are, on average, better**

educated: Holme A, MacArthur C, et al., "The effects of breastfeeding on cognitive and neurological development of children at 9 years." *Child Care Health Dev.* 2010 Jul; 36 (4): 583–90.

74 **a careful study published in July 2013:** Belfort MB, Rifas-Shiman SL, et al., "Infant feeding and childhood cognition at ages 3 and 7 years: Effects of breastfeeding duration and exclusivity." *JAMA Pediatr.* 2013 Jul 29 [Epub ahead of print].

74 **An editorial accompanying the study:** Christakis DA, "Breastfeeding and cognition: Can IQ tip the scale?" *JAMA Pediatr.* 2013 July 29 [Epub ahead of print].

74 **caffeine improves working memory in middle-aged men:** Klaassen EB, de Groot RH, et al., "The effect of caffeine on working memory load-related brain activation in middle-aged males." *Neuropharmacology.* 2013 Jan; 64: 160–67.

74 **It's not just the caffeine that's beneficial:** Shukitt-Hale B, Miller MG, et al., "Coffee, but not caffeine, has positive effects on cognition and psychomotor behavior in aging." *Age.* 2013 Jan 24 [Epub ahead of print].

74 **the benefits of coffee last more than the couple of hours:** Cao C, Loewenstein DA, et al., "High blood caffeine levels in MCI linked to lack of progression to dementia." *J Alzheimers Dis.* 2012; 30 (3): 559–72.

75 **A 2010 study from a memory clinic in Montreal:** Chertkow H, Whitehead V, et al., "Multilingualism (but not always bilingualism) delays the onset of Alzheimer disease: Evidence from a bilingual community." *Alzheimer Dis Assoc Disord.* 2010 Apr–Jun; 24 (2): 118–25.

76 **A 2009 paper from researchers in Italy:** Kovács AM, Mehler J, "Cognitive gains in 7-month-old bilingual infants." *PNAS.* 2009 Apr 21; 106 (16): 6556–60.

76 **a series of studies by psychologist Ellen Bialystok:** See, for instance, Engel de Abreau PM, Cruz-Santos A, et al., "Bilingualism enriches the poor: Enhanced cognitive control in low-income minority children." *Psychol Sci.* 2012; 23 (11): 1364–71.

76 **one of the largest studies:** Humphrey AD, Valian VV, "Multilingualism and cognitive control: Simon and flanker task performance in monolingual and multilingual young adults." Presented at the Psychonomic Society annual meeting, 2012.

77 **A classic study in 1975:** Spirduso WW, "Reaction and movement time as a function of age and physical activity level." *J Gerontol.* 1975; 30: 435–40.

79 **published in 1992 in the journal *Psychology and Aging*:** Hawkins HL, Kramer AF, et al., "Aging, exercise, and attention." *Psychol Aging.* 1992; 7 (4): 643–53.

79 **a study in the journal *Nature*:** Kramer AF, Hahn S, et al., "Ageing, fitness and neurocognitive function." *Nature.* 1999 Jul 29; 400 (6743): 418–19.

80 **In 2010, he published an fMRI study:** Chaddock L, Erickson KI, et al., "A neuroimaging investigation of the association between aerobic fitness, hippocampal volume, and memory performance in preadolescent children." *Brain Res.* 2010 Oct 28; 1358: 172–83.

80 **Another study by Kramer, published in 2012:** Chaddock L, Erickson KI, "A functional MRI investigation of the association between childhood aerobic fitness and neurocognitive control." *Biol Psychol.* 2012 Jan; 89 (1): 260–68.

80 **researchers at Furman University:** Reed JA, Maslow AL, et al., op. cit.

81 **A 2008 Cochrane review . . . cardiovascular exercise:** Angevaren M, Aufdemkampe G, et al., "Physical activity and enhanced fitness to improve cognitive function in older people without known cognitive impairment." *Cochrane Database Syst Rev.* 2008 Jul 16; (3): CD005381.

82 **a study involving 155 women . . . resistance training:** Liu-Ambrose T, Nagamatsu LS, et al., "Resistance training and executive functions: A 12-month randomized controlled trial." *Arch Intern Med.* 2010; 170 (2): 170–78.

83 **A follow-up study published in 2012:** Nagamatsu LS, Handy TC, et al., "Resistance training promotes cognitive and functional brain plasticity in seniors with probable mild cognitive impairment." *Arch Intern Med.* 2012 Apr 23; 172 (8): 666–68.

84 **"Exercise Is Power":** http://www.youtube.com/watch?v=vG6sJm2d4oc.

85 **Justice Ruth Bader Ginsburg:** Anne E. Marinow, "Personal Trainer Bryant Johnson's Clients Include Two Supreme Court Justices," *Washington Post,* March 19, 2013.

86 **Rauscher and colleagues reported in the October 14, 1993, issue**

of *Nature*: Rauscher FH, Shaw GL, et al., "Music and spatial task performance." *Nature*. 1993; 365: 611.

86 ***The Mozart Effect*, and its sequel:** Don Campbell, *The Mozart Effect: Tapping the Power of Music to Heal the Body, Strengthen the Mind, and Unlock the Creative Spirit* (New York: William Morrow, 1997), and *The Mozart Effect for Children: Awakening Your Child's Mind, Health, and Creativity with Music* (New York: HarperCollins, 2000).

86 **Zell Miller . . . was proposing:** Kevin Sack, "Georgia's Governor Seeks Musical Start for Babies," *New York Times*, January 15, 1998.

86 **Bill Clinton . . . Billy Joel, and Viacom CEO Sumner Redstone:** Anemona Hartocollis, "Clinton Visits to Promote School Renovation and Music," *New York Times*, June 17, 2000.

87 **the journal *Nature* had already published two devastating follow-ups:** Steele KM, Bella SM, et al., "Prelude or requiem for the 'Mozart effect'?" *Nature*. 1999 Aug 26; 400 (6747): 827–28. Chabris CF, "Prelude or requiem for the 'Mozart effect'?" *Nature*. 1999 Aug 26; 400 (6747): 826–27.

87 ***Zero Patience*, a film about AIDS:** Stephen Holden, "Review/Film Festival: A Musical about AIDS Crammed with Ideas," *New York Times*, March 26, 1994.

88 **"Music Lessons Enhance IQ":** Schellenberg EG, "Music lessons enhance IQ." *Psychol Sci*. 2004 Aug; 15 (8): 511–14.

89 **Schellenberg . . . collaborated with Sylvain Moreno:** Moreno S, Bialystok E, et al., "Short-term music training enhances verbal intelligence and executive function." *Psychol Sci*. 2011 Nov; 22 (11): 1425–33.

89 **the Bridge Project:** Victoria Sharp, "The Real Lessons for Children in Studying Music," *Telegraph*, February 27, 2013.

90 **Amishi Jha follows a remarkably hectic schedule:** See www .amishi.com.

90 **Jha's article . . . was the cover story:** Amishi Jha, "Being in the Now," *Scientific American Mind*, March 2013.

91 **After publishing a study in 2005 building on Torkel Klingberg's studies:** Rueda MR, Rothbart MK, et al., "Training, maturation, and genetic influences on the development of executive attention." *PNAS*. 2005 Oct 11; 102 (41): 14931–36.

91 **Integrative Body-Mind Training, or IBMT:** Tang Y-Y, Ma Y, et

al., "Short-term meditation training improves attention and self-regulation." *PNAS*. 2007 Oct 23; 104 (43): 17152–56.

93 **In 2010 they published another study:** Tang Y-Y, Lu Q, et al., "Short-term meditation induces white matter changes in the anterior cingulate." *PNAS*. 2010 Aug 31; 107 (35): 15649–52.

93 **their magnum opus:** Tang Y-Y, Lu Q, et al., "Mechanisms of white matter changes inducted by meditation." *PNAS*. 2012 Jun 26; 109 (26): 10570–74.

Chapter 5: Smart Pills and Thinking Caps

95 **twenty-second annual Neuropharmacology Conference:** All presentations at the New Orleans meeting were published in the January 2013, volume 64, edition of the journal *Neuropharmacology*.

97 **Martha Farah . . . presented one of the first studies:** Ilieva I, Boland J, et al., "Objective and subjective cognitive enhancing effects of mixed amphetamine salts in healthy people." *Neuropharmacology*. 2013 Jan; 64: 496–505.

98 **Sahakian presented encouraging results:** Müller U, Rowe JB, et al., "Effects of modafinil on non-verbal cognition, task enjoyment and creative thinking in healthy volunteers." *Neuropharmacology*. 2013 Jan; 64: 490–95.

100 **Tracey Shors . . . described a study she had carried out in mice:** Curlick DM 2nd, Shors TJ, "Training your brain: Do mental and physical (MAP) training enhance cognition through the process of neurogenesis in the hippocampus?" *Neuropharmacology*. 2013 Jan; 64: 506–14.

103 **a six-year follow-up study of 787 adults:** Alpert HR, Connolly GN, et al., "A prospective cohort study challenging the effectiveness of population-based medical intervention for smoking cessation." *Tob Control*. 2013 Jan; 22 (1): 32–37.

104 **study published in 1966 by Harold Kahn:** Kahn HA, "The Dorn Study of Smoking and Mortality among US Veterans. Report on Eight and One-Half Years of Observations." In *Epidemiological Approaches to the Study of Cancer and Other Chronic Diseases*. Monograph No. 19. National Cancer Institute (Washington, D.C.: US Government Printing Office, 1966), 1–125.

105 **Quik treated rhesus monkeys:** Quik M, Cox H, et al., "Nicotine

reduces levodopa-induced dyskinesias in lesioned monkeys." *Ann Neurol.* 2007 Dec; 62 (6): 588–96.

105 **"significant nicotine-associated improvements":** Newhouse P, Kellar K, et al., "Nicotine treatment of mild cognitive impairment: A 6-month double-blind pilot clinical trial." *Neurology.* 2012 Jan; 78 (2): 91–101.

106 **"nicotine is as addictive as heroin":** Sandra Blakeslee, "Nicotine: Harder to Kick . . . Than Heroin." *New York Times Magazine,* March 29, 1987.

106 **"weak reinforcer":** Villegier AS, Lotfipour S, et al., "Tranylcypromine enhancement of nicotine self-administration." *Neuropharmacology.* 2007 May; 52 (6): 1415–25.

106 **In the past six years, researchers . . . have published over a dozen studies:** See, for instance, Potter AS, Newhouse PA, "Acute nicotine improves cognitive deficits in young adults with attention deficit/hyperactivity disorder." *Pharmacol Biochem Behav.* 2008 Feb; 88 (4): 407–17.

106 **Rusted has published a series of studies:** See, for instance, Rusted J, Ruest T, et al., "Acute effects of nicotine administration during prospective memory, an event related fMRI study." *Neuropsychologia.* 2011 Jul; 49 (9): 2362–68.

108 **studies . . . have shown that tDCS can improve outcomes:** See, for example: Turkeltaub PE, Benjon J, et al., "Left lateralizing transcranial direct current stimulation improves reading efficiency." *Brain Stimul.* 2012 Jul; 5 (3): 201–7. Cohen Kadosh R, Soskic S, et al., "Modulating neuronal activity produces specific and long-lasting changes in numerical competence." *Curr Biol.* 2010 Nov 23; 20 (22): 2016–20. Fregni F, Boggio PS, et al., "Anodal transcranial direct current stimulation of prefrontal cortex enhances working memory." *Exp Brain Res.* 2005 Sep; 166 (1): 23–30.

110 **Hamilton has now published ten studies:** See, for instance, Medina J, Norise C, et al., "Finding the right words: Transcranial magnetic stimulation improves discourse productivity in non-fluent aphasia after stroke." *Aphasiology.* 2012 Sep 1; 26 (9): 1153–68.

111 **"thinking cap":** Fox D, "Brain buzz." *Nature.* 2011 Apr 13; 472: 156–59.

111 **Videos . . . on YouTube showing young men experimenting:**

See, for example, http://www.youtube.com/watch?v=6cD5pi9mxD8.

111 **Olson has already published studies:** Ross LA, McCoy D, et al., "Improved proper name recall by electrical stimulation of the anterior temporal lobes." *Neuropsychologia*. 2010 Oct; 48 (12): 3671–74.

Chapter 6: Boot Camp for My Brain

121 **mindfulness meditation CD:** Jon Kabat-Zinn, "Guided Mindfulness Meditation Series 1." Sounds True. September 1, 2005.

Chapter 7: Are You Smarter Than a Mouse?

132 **Hills, coeditor of a 2012 book:** Peter M. Todd, Thomas T. Hills, Trevor W. Robbins, editors, *Cognitive Search: Evolution, Algorithms, and the Brain* (Cambridge, Mass.: MIT Press, 2012).

132 **"an individual's ability to restrict search":** Hills TT, Dukas R, "The Evolution of Cognitive Search." In PM Todd, TT Hills, TW Robbins, editors, *Cognitive Search: Evolution, Algorithms and the Brain* (Cambridge, Mass.: MIT Press, 2012), 13.

133 **when to run and when to tumble:** Hills TT, Jones MN, et al., "Optimal foraging in semantic memory." *Psychol Rev.* 2012 Apr; 119 (2): 431–40. Hills TT, Todd PM, "Search in external and internal spaces: Evidence for generalized cognitive search processes." *Psychol Sci.* 2008 Aug; 19 (8): 802–8.

134 **dopamine plays a key role:** Hills TT, "Animal foraging and the evolution of goal-directed cognition." *Cogn Sci.* 2006 Jan 2; 30 (1): 3–41.

135 **Hills published a study of the "animal fluency task":** Hills TT, Mata R, et al., "Mechanisms of age-related decline in memory search across the adult life span." *Dev Psychol.* 2013 Apr 15 [Epub ahead of print].

135 **A 2013 study in the journal *Memory and Cognition*:** Unsworth N, Brewer GA, et al., "Working memory capacity and retrieval from long-term memory: The role of controlled search." *Mem Cognit.* 2013 Feb; 41 (2): 242–54.

136 **people's ability to remember their Facebook friends:** Unsworth

N, Spillers GJ, et al., "The role of working memory capacity in autobiographical retrieval: Individual differences in strategic search." *Memory*. 2012; 20 (2): 167–76.

137 **duplication of the entire length of its parents' genome:** See, for instance, Dehal P, Boore JL, "Two rounds of whole genome duplication in the ancestral vertebrate." *PLoS Biol*. 2005 Oct; 2 (10): e314.

138 **the double-doubling of our ancestral genome, came to play an essential role in complex cognition:** Ryan TJ, Kopanitsa MV, et al., "Evolution of GluN2A/B cytoplasmic domains diversified vertebrate synaptic plasticity and behavior." *Nat Neurosci*. 2013 Jan; 16 (1): 25–32. Nithianantharajah J, Komiyama NH, et al., "Synaptic scaffold evolution generated components of vertebrate cognitive complexity." *Nat Neurosci*. 2013 Jan; 16 (1): 16–24.

139 **a variant in the *HMGA2* gene resulted in brain size . . . and IQ:** Stein JL, Medland SE, et al., "Identification of common variants associated with human hippocampal and intracranial volumes." *Nat Genet*. 2012 Apr 15; 44 (5): 552–61.

141 **"Human behavior in the maze paralleled that found in rodents":** Brewer GA, Grunfeld IS, et al., "Working memory in rats and humans." Presented at the Psychonomic Society annual meeting, 2011. http://www.psychonomic.org/pdfs/PScompleteProgram2011.pdf.

144 **described only once in a 1981 paper:** Roberts WA, Dale RHI, "Remembrance of places lasts: Proactive inhibition and patterns of choice in rat spatial memory." *Learning and Motivaton*. 1981 Aug; 12 (3): 261–81.

146 **a general intelligence factor can be discerned in mice:** Kolata S, Light K, Matzel LD, "Domain-specific and domain-general learning factors are expressed in genetically heterogeneous CD-1 mice." *Intelligence*. 2008; 36: 619–29.

146 **those who do better at the working-memory task . . . do better on the reasoning:** Kolata S, Light K, et al., "Variations in working memory capacity predict individual differences in general learning abilities among genetically diverse mice." *Neurobiol Learn Mem*. 2005 Nov; 84 (3): 241–46.

146 **smarter on tests of general cognitive abilities:** Light KR, Kolata S, et al., "Working memory training promotes general cognitive

abilities in genetically heterogenous mice." *Curr Biol.* 2010 Apr 27; 20 (8): 777–82.

146 **less age-related loss of attention and learning abilities:** Matzel LD, Light KR, "Longitudinal attentional engagement rescues mice from age-related cognitive declines and cognitive inflexibility." *Learn Mem.* 2011 Apr 26; 18 (5): 345–56.

Chapter 8: Defenders of the Faith

147 **Alan Baddeley . . . first proposed:** Baddeley AD, Hitch G, "Working Memory." In GH Bower, editor, *The Psychology of Learning and Motivation: Advances in Research and Theory*, vol. 8 (New York: Academic Press, 1974), 47–89.

147 **In a paper published in 1999:** Engle RW, Tuholski SW, et al., "Working memory, short-term memory, and general fluid intelligence: A latent-variable approach." *J Exp Psychol Gen.* 1999 Sep; 128 (3): 309–31.

148 **another influential paper, written by Engle alone:** Engle RW, "Working memory capacity as executive attention." *Curr Dir Psychol Sci.* 2002; 11: 19–23.

152 **A fifth study . . . published in June 2010:** Owen AM, Hampshire A, et al., "Putting brain training to the test." *Nature.* 2010 June 10; 465: 775–78.

153 **One of the first studies to specifically test . . . working-memory training and find no benefit:** Redick TS, Shipstead Z, et al., "No evidence of intelligence improvement after working memory training: A randomized, placebo-controlled study." *J Exp Psychol Gen.* 2013 May; 142 (2): 359–79.

155 **Another paper to find no improvement in intelligence:** Chooi W-T, Thompson LA, "Working memory training does not improve intelligence in healthy young adults." *Intelligence.* 2012 Nov–Dec; 40 (6): 531–42.

157 **The third study to find no benefit:** Thompson TW, Waskom ML, "Failure of working memory training to enhance cognition or intelligence." *PLOS ONE.* 2013 May 22; 8 (5): e63614.

157 **The fourth study showing no benefit:** Smith SP, Stibric M, et al., "Exploring the effectiveness of commercial and custom-built games

for cognitive training." *Computers in Human Behavior*. 2013 Nov; 29 (6): 2388–93.

158 **Engle and two of his Georgia Techies published a paper:** Shipstead Z, Hicks KL, Engle RW, "Cogmed working memory training: Does the evidence support the claims?" *J Appl Res Mem Cogn*. 2012; 1: 185–93.

158 **the next three studies Engle cites**: Holmes J, Gathercole SE, "Working memory deficits can be overcome." 2010, op. cit. Holmes J, Gathercole SE, "Adaptive training leads to sustained enhancement of poor working memory in children." 2009, op. cit. Dahlin KIE, "Effects of working memory training on reading in children with special needs." 2011, op. cit.

159 **A few other of these "failed" studies:** Kronenberger WG, Pisoni DB, et al., "Working memory training for children with cochlear implants: A pilot study." *J Speech Lang Hear Res*. 2011 Aug; 54 (4): 1182–96. Løhaugen GC, Antonsen I, et al., "Computerized working memory training improves function in adolescents born at extremely low birth weight." *J Pediatrics*. 2011; 158: 555–61. McNab F, Varrone A, et al., "Changes in cortical dopamine D1 receptor binding associated with cognitive training." *Science*. 2009; 323: 800–802. Olesen PJ, Westerberg H, et al., "Increased prefrontal and parietal activity after training of working memory." *Nat Neurosci*. 2004 Jan; 7 (1): 75–79.

160 **Susan E. Gathercole . . . coauthored two of the studies that Engle critiqued:** Gathercole SE, Dunning DL, Holmes J, "Cogmed training: Let's be realistic about intervention research." *J Appl Res Mem Cogn*. 2012; 1: 201–3.

160 **The first, a randomized study:** Gibson BS, Kronenberger WG, et al., "Component analysis of simple span vs. complex span adaptive working memory exercises: A randomized, controlled trial." *J Appl Res Mem Cogn*. 2012; 1: 179–84.

161 **The group's second paper:** Gibson BS, Gondoli DM, et al., "The future promise of Cogmed working memory training." *J Appl Res Mem Cogn*. 2012; 1: 214–16.

161 **calling Engle's criticisms "overly pessimistic":** Jaeggi SM, Buschkuehl M, et al., "Cogmed and working memory training—current challenges and the search for underlying mechanisms." *J Appl Res Mem Cogn*. 2012; 1: 211–13.

161 **Klingberg, too, contributed a paper:** Klingberg T, "Is working memory capacity fixed?" *J Appl Res Mem Cogn.* 2012; 1: 194–96.

161 **The only paper . . . that shared Engle's nihilistic views:** Hulme C, Melby-Lervåg M, "Current evidence does not support the claims made for CogMed working memory training." *J Appl Res Mem Cogn.* 2012; 1: 197–200.

162 **Stephanie Bennett's study of twenty-one children with Down syndrome:** Bennett SJ, Holmes J, op. cit.

165 **the "ten-thousand-hour rule":** Malcolm Gladwell, *Outliers: The Story of Success* (New York: Little, Brown, 2008).

167 **practice accounted for only about one-third of people's achievements:** Hambrick DZ, Oswald FL, et al., "Deliberate practice: Is that all it takes to become an expert?" *Intelligence.* Published online May 15, 2013.

168 **a study of 155 Texas hold 'em players:** Meinz EJ, Hambrick DZ, et al., "Roles of domain knowledge and working memory capacity in components of skill in Texas Hold'Em poker." *J Appl Res Mem Cogn.* 2012; 1: 34–40.

Chapter 9: Flowers for Ts65Dn

175 **a mouse model of the disorder had recently been developed:** Reeves RH, Irving NG, et al., "A mouse model for Down syndrome exhibits learning and behavior deficits." *Nat Genet.* 1995 Oct; 11 (2): 177–84.

175 **Costa published one of the first studies:** Clark S, Schwalbe J, et al., "Fluoxetine rescues deficient neurogenesis in hippocampus of the Ts65Dn mouse model for Down syndrome." *Exp Neurol.* 2006 Jul; 200 (1): 156–61.

176 **giving mice with Down syndrome the Alzheimer's drug memantine:** Costa AC, Scott-McKean JJ, et al., "Acute injections of the NMDA receptor antagonist memantine rescue performance deficits of the Ts65Dn mouse model of Down syndrome on a fear conditioning test." *Neuropsychopharmacology.* 2008 Jun; 33 (7): 1624–32.

177 **John Langdon H. Down was the first to describe the disorder:** Down JLH, "Observations on an ethnic classification of idiots." *London Hospital Reports.* 1866; 3: 259–62.

177 **the disorder is caused by a third copy of the 21st chromosome:** Lejeune J, Turpin R, et al., "Mongolism: A chromosomal disease (trisomy)." *Bull Acad Natl Med.* 1959 Apr 7–14; 143 (11–12): 256–65.

180 **William C. Mobley . . . coauthored a study:** Salehi A, Faizi M, et al., "Restoration of norepinephrine-modulated contextual memory in a mouse model of Down syndrome." *Sci Transl Med.* 2009 Nov 18; 1 (7): 7ra17.

180 **Paul Greengard . . . entered the fray:** Netzer WJ, Powell C, et al., "Lowering beta-amyloid levels rescues learning and memory in a Down syndrome mouse model." *PLoS One.* 2010 Jun 3; 5 (6): e10943.

181 ***Neurology Today* . . . article about Mobley's mouse study:** Dan Hurley, "Drug study is latest to show improved cognition in mouse model for Down syndrome." *Neurol Today.* 2010 Jan 7; 10 (1): 1, 11–12.

190 **Kennedy called for immigration law changes:** "Sen. Kennedy Urges Immigration Law Check," *Lodi News-Sentinel*, June 8, 1971, p. 9.

192 **A 2009 survey conducted in Canada:** Karen Kaplan, "Is a Wonder Pill Necessarily Wonderful for People with Down Syndrome?" *Los Angeles Times*, November 18, 2009, http://latimesblogs.latimes.com/booster_shots/2009/11/down-syndrome-treatment.html.

192 ***Life As We Know It*:** Michael Bérubé, *Life As We Know It: A Father, a Family, and an Exceptional Child* (New York: Pantheon, 1996).

193 **Released online by the journal *Translational Psychiatry*:** Boada R, Hutaff-Lee C, et al., "Antagonism of NMDA receptors as a potential treatment for Down syndrome: A pilot randomized controlled trial." *Transl Psychiatry.* 2012 Jul 17; 2: e141.

Chapter 10: Clash of the Titans

197 **Charles Hulme and Monica Melby-Lervåg . . . published a new meta-analysis:** Hulme C, Melby-Lervåg M, "Is working memory training effective? A meta-analytic review." *Dev Psychol.* 2013 Feb; 49 (2): 270–91. Epub 2012 May 21.

198 **a short blog item posted online:** Gareth Cook, "Brain Games Are Bogus," *New Yorker*, April 5, 2013, http://www.newyorker.com/online/blogs/elements/2013/04/brain-games-are-bogus.html.

199 **a second meta-analysis:** Hindin SB, Zelinski EM, "Extended practice and aerobic exercise interventions benefit untrained cognitive outcomes in older adults: A meta-analysis." *J Am Geriatr Soc.* 2012 Jan; 60 (1) 136–41.

199 **two *other* systematic reviews of the literature:** Kueider AM, Parisi JM, et al., "Computerized cognitive training with older adults: A systematic review." *PLoS One.* 2012; 7 (7): e40588. Reijnders J, van Heugten C, et al., "Cognitive interventions in healthy older adults and people with mild cognitive impairment: A systematic review." *Ageing Res Rev.* 2013 Jan; 12 (1): 263–75.

209 **"improvement in working memory . . . transfer[s] to second-language learning":** Phillips AM, Eddington CM, et al., "Far transfer from adaptive cognitive training to L2 vocabulary learning." Poster 5118 presented at the fifty-third annual meeting of the Psychonomic Society, Minneapolis, Minnesota, 2012 Nov 17.

210 **Jensen . . . published a paper in the *Harvard Educational Review*:** Jensen AR, "How much can we boost IQ and scholastic achievement?" *Harvard Educational Review.* 1969 Winter; 39 (1): 1–123.

210 **the *New York Times* printed an op-ed column:** Nicholas Kristof, "It's a Smart, Smart, Smart World," *New York Times*, December 12, 2012, p. A35.

211 **the Flynn effect:** See, for instance, James R. Flynn, *What Is Intelligence?* (New York: Cambridge University Press, 2007) and *Are We Getting Smarter? Rising IQ in the Twenty-First Century* (New York: Cambridge University Press, 2012).

213 **"Venezuela Intelligence Project":** See, for example, Nickerson RS, "Project Intelligence: An account and some reflections." *Special Services in the Schools.* 1986 Fall–Winter; 3 (1): 83–102; and Herrnstein EJ, Nickerson RS, et al., op. cit.

213 **opposes "the teaching of Higher Order Thinking Skills":** 2012 Texas State Republican Party Platform, p. 12. http://s3.amazonaws.com/texasgop_pre/assets/original/2012Platform_Final.pdf.

INDEX